T0258285

Concepts and Functions of Ferromagnetic Resonance

Concepts and Functions of Ferromagnetic Resonance

Edited by **Sharon Tatum**

New York

Published by NY Research Press,
23 West, 55th Street, Suite 816,
New York, NY 10019, USA
www.nyresearchpress.com

Concepts and Functions of Ferromagnetic Resonance
Edited by Sharon Tatum

International Standard Book Number: 978-1-63238-096-8 (Hardback)

Printed in the United States of America.

Contents

Preface

The main aim of this book is to educate learners and enhance their research focus by presenting diverse topics covering this vast field. This is an advanced book which compiles significant studies by distinguished experts in the area of analysis. This book addresses successive solutions to the challenges arising in the area of application, along with it; the book provides scope for future developments.

The book offers an in-depth look at ferromagnetic resonance, demonstrating its functions and concepts. It lays emphasis on current developments in the interaction of science and technology of nano-structures like nano-wires, photonic crystals, spinel type nano-particles etc. Electromagnetic resonance techniques now play a pivotal role in science and technology. The method of ferromagnetic resonance is most vastly used in sprintronics, electronics, space navigators, remote control equipment, radio engineering, electronic computers, maritime electrical engineering, instrument making and geophysical programs of prospecting.

It was a great honour to edit this book, though there were challenges, as it involved a lot of communication and networking between me and the editorial team. However, the end result was this all-inclusive book covering diverse themes in the field.

Finally, it is important to acknowledge the efforts of the contributors for their excellent chapters, through which a wide variety of issues have been addressed. I would also like to thank my colleagues for their valuable feedback during the making of this book.

Editor

Ferromagnetic Resonance

Orhan Yalçın

Additional information is available at the end of the chapter

1. Introduction

Ferromagnetism is used to characterize magnetic behavior of a material, such as the strong attraction to a permanent magnet. The origin of this strong magnetism is the presence of a spontaneous magnetization which is produced by a parallel alignment of spins. Instead of a parallel alignment of all the spins, there can be an anti-parallel alignment of unequal spins. This results in a spontaneous magnetization which is called ferrimagnetism.

The resonance arises when the energy levels of a quantized system of electronic or nuclear moments are Zeeman split by a uniform magnetic field and the system absorbs energy from an oscillating magnetic field at sharply defined frequencies corresponding to the transitions between the levels. Classically, the resonance event occurs when a transverse ac field is applied at the Larmor frequency.

The resonance behaviour usually called magnetic resonance (MR) and nuclear magnetic resonance (NMR). Main types of resonance phenomenon can be listed as nuclear magnetic resonance (NMR), nuclear quadrupole resonance (NQR), electron paramagnetic/spin resonance (EPR, ESR), spin wave resonance (SWR), ferromagnetic resonance (FMR), antiferromagnetic resonance (AFMR) and conductor electron spin resonance (CESR). The resonant may be an isolated ionic spin as in electron paramagnetic resonance (EPR) or a nuclear magnetic resonance (NMR). Also, resonance effects are associated with the spin waves and the domain walls. The resonance methods are important for investigating the structure and magnetic properties of solids and other materials. These methods are used for imaging and other applications.

The following information can be accessed with the help of such resonance experiments. (i) Electrical structure of point defects by looking at the absorption in a thin structure. (ii) The line width with the movement of spin or surroundings isn't changed. (iii) The distribution of the magnetic field in solid by looking at the of the resonance line position (chemical shift and etc.). (iv) Collective spin excitations.

The atoms of ferromagnetic coupling originate from the spins of d-electrons. The size of μ permanent atomic dipoles create spontaneously magnetized. According to the shape of dipoles materials can be ferromagnetic, antiferromagnetic, diamagnetic, paramagnetic and etc.

Ferromagnetic resonance (FMR) technique was initially applied to ferromagnetic materials, all magnetic materials and unpaired electron systems. Basically, it is analogous to the electron paramagnetic resonance (EPR). The EPR technique gives better results at unpaired electron systems. The FMR technique depends on the geometry of the sample at hand. The demagnetization field is observed where the sample geometry is active. The resonance area of the sample depends on the properties of material. The FMR technique is advantageous because it does not cause damage to materials. Also, it allows a three dimensional analysis of samples. The FMR occurs at high field values while EPR occurs at low magnetic field values. Also, line-width of ferromagnetic materials is large according to paramagnetic materials. Exchange interaction energy between unpaired electron spins that contribute to the ferromagnetism causes the line narrowing. So, ferromagnetic resonance lines appear sharper than expected.

The FMR studies have been increased since the EPR was discovered in 1945 (Zavosky, 1945; Kittel, 1946, 1947, 1949, 1953, 1958; Kip, 1949; Bloembergen, 1950, 1954; Crittenden, 1953; Van Vleck, 1950; Herring, 1950; Anderson, 1953; Damon, 1953; Young, 1953; Ament, 1955; Ruderman, 1954; Reich, 1955; Kasuya, 1956; White, 1956; Macdonald, 1956; Mercereau, 1956; Walker, 1957; Yosida, 1957; Tannenwald, 1957; Jarrett, 1958; Rado, 1958; Brown, 1962; Frait, 1965; Sparks, 1969). The beginnings of theoretical and experimental studies of spectroscopic investigations of basic sciences are used such as physics, chemistry, especially nanosciences and nanostructures (Rodbell, 1964; Kooi, 1964; Bhagat, 1967, 1974; Sparks, 1970(a), 1970(b), 1970(c), 1970(d); Rachford, 1981; Dillon, 1981; Schultz, 1983; Artman, 1957, 1979; Ramesh, 1988(a), 1988(b); Fraitova, 1983(a), 1983(b), 1984; Teale, 1986; Speriosu, 1987; Vounyuk, 1991; Roy, 1992; Puszkarski, 1992; Weiss, 1955). The FMR technique can provide information on the magnetization, magnetic anisotropy, dynamic exchange/dipolar energies and relaxation times, as well as the damping in the magnetization dynamics (Wigen, 1962, 1984, 1998; De Wames, 1970; Wolfram, 1971; Yu, 1975; Frait, 1985, 1998; Rook, 1991; Bland, 1994; Patton, 1995, 1996; Skomski, 2008; Coey, 2009). This spectroscopic method/FMR have been used to magnetic properties (Celinski, 1991; Farle, 1998, 2000; Fermin, 1999; Buschow, 2004; Heinrich, 2005(a), 2005(b)), films (Özdemir, 1996, 1997), monolayers (Zakeri, 2006), ultrathin and multilayers films (Layadi, 1990(a), 1990(b), 2002, 2004; Wigen, 1993; Zhang, 1994(a), 1994(b); Farle, 2000; Platow, 1998; Anisimov, 1999; Yıldız, 2004; Heinrich, 2005(a); Lacheisserie, 2005; de Cos, 2006; Liua, 2012; Schäfer, 2012), the angular, the frequency (Celinski, 1997; Farle, 1998), the temperature dependence (Platow, 1998), interlayer exchange coupling (Frait, 1965, 1998; Parkin, 1990, 1991(a), 1991(b), 1994; Schreiber, 1996; Rook, 1991; Wigen, 1993; Layadi, 1990(a); Heinrich, 2005; Paul, 2005), Brillouin light scattering (BLS) (Grünberg, 1982; Cochran, 1995; Hillebrands, 2000) and sample inhomogeneities (Artman, 1957, 1979; Damon, 1963; McMichael, 1990; Arias, 1999; Wigen, 1998; Chappert, 1986; Gnatzig, 1987; Fermin, 1999) of samples. Besides using FMR to characterize magnetic properties, it also allows one to study the fundamental excitations and technological applications of a magnetic system (Schmool, 1998; Voges, 1998; Zianni, 1998; Grünberg, 2000, 2001; Vlasko-Vlasov, 2001; Zhai, 2003; Aktaş, 2004; Birkhäuser Verlag, 2007; Seib, 2009). The various thickness, disk array, half-metallic ferromagnetic electrodes, magnon scattering and other of some properties of samples have been studied using the FMR tehniques (Mazur, 1982; da Silva, 1993; Chikazumi, 1997; Song, 2003; Mills, 2003; Rameev, 2003(a), 2003(b), 2004(a), 2004(b); An, 2004; Ramprasad, 2004; Xu, 2004; Wojtowicz, 2005; Zakeri, 2007; Tsai, 2009; Chen, 2009). The magnetic properties of single-crystalline (Kambe, 2005; Brustolon, 2009), polycrystalline (Singh, 2006; Fan, 2010), alloy films (Sihues, 2007), temperature dependence and similar qualities have been studied electromagnetic spectroscopy techniques (Özdemir, 1998; Birlikseven, 1999(a), 1999(b); Fermin, 1999; Rameev, 2000; Aktaş, 2001; Budak, 2003; Khaibullin, 2004). The magnetic resonance techniques (EPR, FMR) have been applied to the iron oxides, permalloy nanostructure (Kuanr, 2005), clustered, thermocouple connected to the ferromagnet, thin permalloy layer and et al. (Guimarães, 1998; Spoddig, 2005; Can, 2012; Rousseau, 2012; Valenzuela, 2012; Bakker, 2012; Maciá, 2012; Dreher, 2012; Kind, 2012; Li, 2012; Estévez, 2012; Sun, 2012(a), 2012(b), 2012(c); Richard, 2012). Magneto optic (Paz, 2012), dipolar energy contributions (Bose, 2012), nanocrystalline (Maklakov, 2012; Raita, 2012), $La_{0.7}Sr_{0.3}MnO_3$ films (Golosovsky, 2012), $La_{0.67}Ba_{0.33}Mn_{1-y}A_yO_3$, A - Fe, Cr (Osthöver, 1998), voltage-controlled magnetic anisotropy (VCMA) and spin transfer torque (Zhu, 2012) and the typical properties of the inertial resonance are investigated (Olive, 2012). The exchange bias (Backes, 2012), Q cavities for magnetic material (Beguhn, 2012), MgO/CoFeB/Ta structure (Chen, 2012), the interfacial origin of the giant magnetoresistive effect (GMR) phenomenon (Prieto, 2012), self-demagnetization field (Hinata, 2012), $Fe_3O_4/InAs(100)$ hybrid spintronic structures (Huang, 2012), granular films (Kakazei, 1999, 2001; Sarmiento, 2007; Krone, 2011; Kobayashi, 2012), nano-sized powdered barium ($BaFe_{12}O_{19}$) and strontium (Sr $Fe_{12}O_{19}$) hexaferrites (Korolev, 2012), $Ni_{0.7}Mn_{0.3-x}Co_xFe_2O_4$ ferrites (NiMnCo: x = 0.00, 0.04, 0.06, and 0.10) (Lee, 2012), thin films (Demokritov, 1996,1997; Nakai, 2002; Lindner, 2004; Aswal, 2005; Jalali-Roudsar, 2005; Cochran, 2006; Mizukami, 2007; Seemann, 2010), Ni_2MnGa films (Huang, 2004), magnetic/electronic order of films (Shames, 2012), $Fe_{1-x}Gd(Tb)_x$ films (Sun, 2012), in ε-$Al_{0.06}Fe_{1.94}O_3$ (Yoshikiyo, 2012). 10 nm thick Fc/GaAs(110) film (Römer, 2012), triangular shaped permalloy rings (Ding, 2012) and Co_2-Y hexagonal ferrite single rod (Bai, 2012) structures and properties have been studied by FMR tecniques (Spaldin, 2010). Biological applications (Berliner, 1981; Wallis, 2005; Gatteschi, 2006; Kopp, 2006; Fischer, 2008; Mastrogiacomo, 2010), giant magneto-impedance (Valenzuela, 2007; Park, 2007), dynamics of feromagnets (Vilasi, 2001; Rusek, 2004; Limmer, 2006; Sellmyer, 2006; Spinu, 2006; Azzerboni, 2006; Krivoruchko, 2012), magneto-optic kerr effect (Suzuki, 1997; Neudecker, 2006), Heusler alloy (HA) films (Kudryavtsev, 2007), ferrites (Kohmoto, 2007), spin polarized electrons (Rahman, 2008) and quantum mechanics (Weil, 2007) have been studied by FMR technique in generally (Hillebrands, 2002, 2003, 2006). In additional, electric and magnetic properties of pure, Cu^{2+} ions doped hydrogels have been studied by ESR techniques (Coşkun, 2012).

The FMR measurements were performed in single crystals of silicon- iron, nickel-iron, nickel and hcp cobalt (Frait, 1965), thin films (Knorr, 1959; Davis, 1965; Hsia, 1981; Krebs, 1982; Maksymowich, 1983, 1985, 1992; Platow, 1998; Durusoy, 2000; Baek, 2002; Kuanr, 2004), CoCr magnetic thin films (Cofield, 1987), NiFe/FeMn thin films (Layadi, 1988), single-crystal Fe/Cr/Fe(100) sandwiches (Krebs, 1989), polycrystalline single films (Hathaway, 1981; Rezende, 1993) and ultrathin multilayers of the system Au/Fe/Au/Pd/Fe (001) prepared on GaAs(001) (Woltersdorf, 2004). The FMR techniques have been succesfully applied peak-to-peak linewidth (Yeh, 2009; Sun, 2012), superconducting and ferromagnetic coupled structures (Richard, 2012) and thin Co films of 50 nm thick (Maklakov, 2012). The garnet materials (Ramesh, 1988 (a), 1988 (b)), polar magneto-optic kerr effect and brillouin light scattering measurements (Riedling, 1999), giant-magnetoresistive (GMR) multilayers (Grünberg, 1991; Borchers, 1998) and insulated multilayer film (de Cos, 2006; Lacheisserie, 2005) are the most intensely studied systems.

The technique of FMR can be applied to nano-systems (Poole, 2003; Parvatheeswara, 2006; Mills, 2006; Schmool, 2007; Vargas, 2007; Seemann, 2009; Wang, 2011; Patel, 2012; De Biasi, 2013). The FMR measurement on a square array of permalloy nanodots have been comparion a numerical simulation based on the eigenvalues of the linearized Landau-Lifshitz equation (Rivkin, 2007). The dynamic fluctuations of the nanoparticles and their anisotropic behaviour have been recorded with FMR signal (Owens, 2009). Ferromagnetic resonance (FMR) modes for $Fe_{70}Co_{30}$ magnetic nanodots of 100 nm in diameter in a mono-domain state are studied under different in-plane and out-of-plane magnetic fields (Miyake, 2012). The FMR techniques have been accomplished applied to magnetic microwires and nanowire arrays (Adeyeye, 1997; Wegrowe, 1999, 2000; García-Miquel, 2001; Jung, 2002; Arias, 2003; Raposo, 2011; Boulle, 2011; Kraus, 2012; Klein, 2012). In additional, FMR measurements have been performed for nanocomposite samples of varying particles packing fractions with demagnetization field (Song, 2012). The ferromagnetic resonance of magnetic fluids were theoretically investigated on thermal and particles size distribution effects (Marin, 2006). The FMR applied to nanoparticles, superparamagnetic particles and catalyst particles (de Biasi, 2006; Vargas, 2007; Duraia, 2009).

In the scope of this chapter, we firstly give a detailed account of both magnetic order and their origin. The origin of magnetic orders are explained and the equations are obtained using Fig.1 which shows rotating one electron on the table plane. Then, the dynamic equation of motion for magnetization was derived. We mentioned MR and damping terms which have consisted three terms as the Bloch-Bloembergen, the Landau-Lifshitz and the Gilbert form. We indicated electron EPR/ESR and their historical development. The information of spin Hamiltonian and g-tensor is given. The dispersion relations of monolayer, trilayers, five-layers and multilayer/n-layers have regularly been calculated for ferromagnetic exchange-couple systems (Grünberg, 1992; Nagamine, 2005, Schmool, 1998). The theoretical FMR spectra were obtained by using the dynamic equation of motion for magnetization with the Bloch-Bloembergen type damping term. The exchange-spring (hard/soft) system which is the best of the sample for multilayer structure has been explained by using the FMR technique and equilibrium condition of energy of system. The

FMR spectra originated from the iron/soft layers as shown in the exchange spring magnets in Fig.9. Finally, superparamagnetic/single-domain nanoparticles and their resonance are described in detail.

2. Magnetic order

Magnetic materials are classified as paramagnetic, ferromagnetic, ferrimagnetic, antiferromagnetic and diamagnetic to their electronic order. Magnetic orders are divided in two groups as (i) paramagnetic, ferromagnetic, ferrimagnetic, antiferromagnetic and (ii) diamagnetic. The magnetic moments in diamagnetic materials are opposite to each other as well as the moments associated with the orbiting electrons so that a zero magnetic moment (μ) is produced on macroscopic scale. In the paramagnetic materials, each atom possesses a small magnetic moment. The orientation of magnetic moment of each atom is random, the net magnetic moment of a large sample (macroscopic scale) of dipole and the magnetization vector are zero when there is no applied field.

Nanoscience, nanotechnology and nanomaterials have become a central field of scientific and technical activity. Over the last years the interest in magnetic nanostructures and their applications in various electronic devices, effective opto-electronic devices, bio-sensors, photo-detectors, solar cells, nanodevices and plasmonic structures have been increasing tremendously. This is caused by the unique properties of magnetic nanostructures and the outstanding performance of nanoscale devices. Dimension in the range of one to hundred nanometers, is called the nano regime. In recent years, nanorods, nanoparticles, quantum dots, nanocrystals etc. are in a class of nanostructures (Yalçın, 2012; Kartopu & Yalçın, 2010; Aktaş, 2006) studied extensively. As the dimensions of nano materials decrease down to the nanometer scale, the surface of nanostructures starts to exhibit new and interesting properties mainly due to quantum size effects.

3. Origin of magnetic moment

The magnetization of a matter is derived by electrons moving around the nucleus of an atom. Total magnetic moment occurs when the electrons such as a disc returns around its axis consist of spin angular momentum and returns around the nucleus consist orbital angular momentum. The most of matters which have unpaired electrons have a little magnetic moment. This natural angular momentum consists of the result of charged particle return around its own axis and is called spin of the particle. The origin of spin is not known exactly, although electron is point particle the movement of an electron in an external magnetic field is similar to the movement of the disc. In other words, the origin of the spin is quantum field theoretical considerations and comes from the representations of the Poincare algebra for the elementary particles. The magnetism related to spin angular momentum, orbital angular momentum and spin-orbit interactions angular momentum. The movement of the electron around the nucleus can be considered as a current loop while electron spin is considered very small current loop which generate magnetic field. Here, orbital angular momentum was obtained by the result of an electron current loop around the nucleus. Thus, both it is exceeded the difficulty of understanding the magnetic moment and the magnetic moment for an electron orbiting around the nucleus is used easily. The result of orbital-angular momentum ($\vec{L}$) adapted for spin-angular momentum ($\vec{S}$) (Cullity, 1990).

One electron is rotating from left to right on the table plane as shown in Fig.1. The rotating electron creates a current (i) on the circle with radius of r.

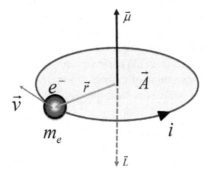

Figure 1. Schematic representation of the precession of a single electron on the table plane.

The magnetic moment of a single electron is defined as below

$$\vec{\mu} = i \cdot \vec{A}. \tag{1}$$

Where, $\vec{A}$ is the circle area. The magnetic moment is written as follows by using the current ($-e = i \cdot t$), one cycle ($2\pi r = v \cdot t$) and angular momentum ($L = m_e \cdot v \cdot r$) definition.

$$\vec{\mu} = -\frac{e}{2m_e} \vec{L} \tag{2}$$

Where $\gamma = e / 2m_e$ and $\vec{L}$ is the gyromagnetic (magneto-mechanical or magneto-gyric) ratio and the orbital-angular momentum, respectively. Therefore, the magnetic moment $\vec{\mu}$ is obtained from Eq. (2) as below

$$\vec{\mu} = -\gamma \vec{L}. \tag{3}$$

The following expression is obtained when derivative of Eq.(3)

$$d\vec{\mu} + \gamma \, d\vec{L} + \vec{L} d\gamma = 0. \tag{4}$$

For our purpose, we only need to know that γ is a constant and $d\gamma = 0$. From this results, $d\vec{\mu} + \gamma \, d\vec{L} = 0$. The derivative of time of this equation, the equation of motion for magnetic moments of an electron is found as below

$$\frac{1}{\gamma}\frac{d\vec{\mu}}{dt} = \frac{d\vec{L}}{dt} = \vec{\tau}. \tag{5}$$

This equation is related to $\vec{\tau} = d\vec{L}/dt$ in two dimensional motions on the plane and $\vec{F} = d\vec{P}/dt$ in one dimensional motion. This motion corresponds to Newton's dynamic equations. When an electron is placed in an applied magnetic field $\vec{H}$, the magnetic field will produce a torque ($\vec{\tau}$) on the magnetic moment ($\vec{\mu}$) of amount $\vec{\mu} \times \vec{H}$. The equation of motion for magnetic moment ($\vec{\mu}$) is found by equating the torque as below

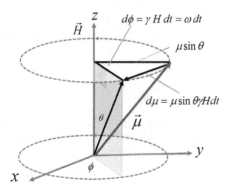

Figure 2. Schematic representation of precession of a single magnetic moment $\vec{\mu}$ in the external magnetic field around the z-axis.

$$\frac{1}{\gamma}\frac{d\vec{\mu}}{dt} = \vec{\mu} \times \vec{H}. \tag{6}$$

This expression is called the equation of motion for magnetic moment ($\vec{\mu}$). The motion of magnetic moment ($\vec{\mu}$) forms a cone related to $\vec{H}$ when the angle θ of magnetic moment and external magnetic field does not change. Therefore, in time (dt), the tip of the vector $\vec{\mu}$ moves an angle $(\gamma H)dt$. The magnetic moment vector make precession movement about $\vec{H}$ at a frequency of $\gamma H/2\pi$. This frequency, $\upsilon = \omega/2\pi = \gamma H/2\pi$, is called the Larmor frequency. In general this Larmor frequency is used this form $\omega = \gamma H$ in literature.

4. Magnetic resonance

Magnetic Resonance (MR) is a research branch which examines magnetic properties of matters. The magnetic properties of atom originate from electrons and nucleus. So, it is studied in two groups such as electron paramagnetic resonance (EPR)/electron spin resonance (ESR) and nuclear magnetic resonance (NMR). At ESR and NMR all of them are the sample is placed in a strong static magnetic field and subjected to an orthogonally amplitude-frequency. While EPR uses a radiation of microwave frequency in general, NMR is observed at low radio frequency range. The energy absorption occurs when radio frequency is equal with energy difference between electrons two levels. But, the transition must obey the selection rules. The splitting between the energy levels occurs when total angular moment of electron is different from zero. On the other hand, the splitting of energy levels has not been observed in the filled orbit. The precession motion of a paramagnetic sample in magnetic field is seen schematically in Fig. 2. If microwave field with υ-frequency at perpendicular is applied to the static field, it comes out power absorption when precession (ω_0) is same with υ-frequency. The power increases when these frequencies come near to each other and it occurs maximum occurs at point when they are equal. This behaviour is called magnetic resonance (MR).

The magnetic materials contain a large number of atomic magnetic moment in generally. Net atomic magnetic moment can be calculated by $\vec{M} = N\vec{\mu}$. Where, N is the number of atomic magnetic moment in materials.

$$\frac{1}{\gamma}\frac{d\vec{M}}{dt} = \vec{M} \times \vec{H}_{eff} \tag{7}$$

This precession movement continue indefinitely would take forever when there is no damping force. The damping term may be introduced in different ways. Indeed, since the details of the damping mechanism in a ferromagnet have not been completely resolved, different mathematical forms for the damping have been suggested. The three most common damping terms used to augment the right-hand side of Eq. (7) are as follows:

(i) The Bloch-Bloembergen form: $-\dfrac{\vec{M}_{\theta,\varphi}}{T_2} - \dfrac{\vec{M}_z - M_0}{T_1}$

(ii) The Landau-Lifshitz form: $\dfrac{-\lambda}{\left|\vec{M}\right|^2}\vec{M} \times \vec{M} \times \vec{H}$

(iii) The Gilbert form: $\dfrac{\alpha}{\left|\vec{M}\right|}\vec{M} \times \dfrac{d\vec{M}}{dt}$

Bloch-Bloembergen type damping does not converse M so it is equivalent to the type of Landau-Lifshitz and the Gilbert only when α is small and for small excursion of $\vec{M}$. For large excursion of M, the magnitude of $\vec{M}$ is certainly not protected, as the damping torque is in the direction of the magnetization component in this formularization. Hence, the observation of M in the switching experiments in thin films should be provide a sensitive test on the appropriate form of the damping term for ferromagnetism since $\vec{M}$ which is conserved during switching. This would suggest that the form of the Bloch-Bloembergen damping term would not be applicable for this type of experiment. The Gilbert type (Gilbert, 1955) is essentially a modification of the original form which is proposed firstly by Landau and Lifshitz (Landau & Lifshitz, 1935). It is very important to note that the Landau-Lifshitz and Gilbert type of damping conserve while the Bloch-Bloembergen (Bloembergen,

1950) type does not. Landau and Lifshitz observed that the ferromagnetic exchange forces between spins are much greater than the Zeeman forces between the spins and the magnetic fields in their formulation of the damping term. Therefore, the exchange will conserve the magnitude of $\vec{M}$. In this formulation, since the approach of $\vec{M}$ towards $\vec{H}$ is due completely to the relatively weak interaction between $\vec{M}$ and $\vec{H}$, we must require that $\lambda << \gamma M$. In this small damping limit, the Landau-Lifshitz and the Gilbert forms are equivalence so that whether one uses one or the other is simply a matter of convenience or familiarity. However, Callen has obtained a dynamic equation by quantizing the spin waves into magnons and treating the problem quantum-mechanically (Callen, 1958). Subsequently, Fletcher, Le Craw, and Spencer have reproduced the same equation using energy consideration (Fletcher, 1960). In their reproduction, they found the mean the rate of energy transfer between the uniform precession, the spin waves (Grünberg, 1979, 1980) and the lattice.

5. Electron paramagnetic resonance

Stern and Gerlach (Gerlach, 1922) proved that the electron-magnetic moment of an atom in an external magnetic field originates only in certain directions in the experiment in 1922. Uhlenbek and Goudsmit found that the connection between the magnetic moment and spin angular momentum of electron (Uhlenbek, 1925), Rabi and Breit found the transition between the energy levels in oscillating magnetic field (Rabi, 1938). This also proved to be observed in the event of the first magnetic resonance. The EPR technique is said to be important of Stern-Gerlach experiment. Zavoisky observed the first peak in the electron paramagnetic resonance for $CuCl_22H_2O$ sample and recorded (Zavoisky, 1945). The most of EPR experiments were made by scientists in the United Kingdom and the United States. Important people mentioned in the experimental EPR studies; Abragam, Bleaney and Van Vleck. The historical developments of MR have been summarized by Ramsey (Ramsey, 1985). NMR experiments had been done by Purcell et al. (Purcell, 1946). Today it has been used as a tool for clinical medicine. MRI was considered as a basic tool of CT scan in 1970s. The behaviors of spin system under the external magnetic field with the gradient of spin system are known NMR tomography. This technique is used too much for medicine, clinics, diagnostic and therapeutic purposes. General structure of the EPR spectrometer consist four basic parts in general. (i) Source system (generally used in the microwave 1-100 GHz), (ii) cavity-grid system, (iii) Magnet system and (iv) detector and modulation system. EPR/ESR is subject of the MR. An atom which has free electron when it is put in magnetic field the electron's energy levels separate (Yalçın, 2003, 2007(a), 2007(b)). This separation originates from the interaction of the electrons magnetic moment with external magnetic field. Energy separating has been calculated by the following Hamiltonian.

$$\hat{H} = g\,\mu_B\,\vec{H} \cdot \hat{S} \tag{8}$$

It is called Zeeman Effect. If the applied magnetic field oriented z-axis energy levels are;

$$E_{Ms} = g\,\mu_B\,H \cdot M_s. \tag{9}$$

Here, g is the g-value (or Landé g-value) (for free electron $g_e = 2.0023193$ and proton $g_N = 2.7896$), μ_B is Bohr magneton ($\mu_B = \left(eh/4\pi m_e\right) = 9.2740 \times 10^{-24} \, J/T$) and M_s is the number of magnetic spin quantum. If the orbital angular momentum of electron is large of zero ($L > 0$) g -value for free atoms is following

$$g = 1 + \frac{S(S+1) - L(L+1) + J(J+1)}{2J(J+1)}. \tag{10}$$

The anisotropy of the g-factor is described by taking into account the spin–orbit interaction combined (Yalçın, 2004(c)). The total magnetic moment can be written at below;

$$\mu_{eff} = g\,\mu_B \sqrt{J(J+1)}$$

The values of orbital angular momentum of unpaired electrons for most of the radicals and radical ions are zero or nearly zero. Hence, the number of total electron angular momentum J equals only the number of spin quantum S. So, these values are nearly 2. For free electron ($M_s = \pm 1/2$) and for this electron;

$$\Delta E = E_{+1/2} - E_{-1/2} = g\,\mu_B\,H. \tag{11}$$

When the electromagnetic radiation which frequency υ is applied to such an electron system;

$$h\upsilon = g\,\mu_B\,H. \tag{12}$$

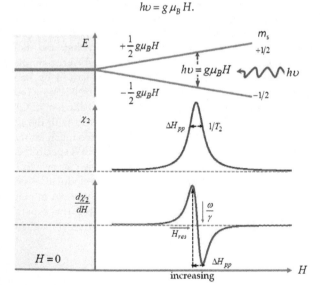

Figure 3. The energy levels and resonance of free electron at zero field and increasing applied magnetic field. In this figure, while the value of magnetic field increases, the separating between energy levels increase. Arbitrary units used in vertical axes for χ_2 and $d\chi_2/dH$.

If this equation is provided the system absorbs energy from applied electromagnetic wave (see Fig. 3). It is called resonance effect. Material absorbs energy in two different ways from applied electromagnetic wave by according to the Eq.(12). Firstly, in Eq.(12), frequency of electromagnetic wave doesn't change while the external magnetic field changes. Secondly, its opposite can be provided.

In this Fig.(3) it has been seen that magnetic susceptibility χ_2 versus magnetic field. At the same time it is said the absorption curve. The magnetic field derivative beneath of this figure is FMR absorption spectrum ($d\chi_2/dH$). Here, Δ_{pp} and $1/T_2$ are linewidth, H_{res} is resonance field, ω/γ is resonance frequency.

5.1. Spin Hamiltonian

The spin Hamiltonian is total electronic spins and nucleon spin $\vec{I}$ which have crystal lattice under the static magnetic field following;

$$\hat{H} = \mu_B \vec{H} \cdot \vec{g} \cdot \hat{S} + \hat{S} \cdot \vec{D} \cdot \hat{S} + \vec{S} \cdot \vec{A} \cdot \hat{I} - \mu_N \cdot \vec{H} \cdot \vec{g}_N \cdot \hat{I} + \hat{I} \cdot \vec{P} \cdot \hat{I} \tag{13}$$

$\hat{S}$ and $\vec{I}$ operators of electronic and nucleus, respectively. In this equation, first term is Zeeman effect, second one is thin layer effects, the third one is the effect of between electronic spin and nucleus-spin of ion and it is known that thin layer effects. The fourth term is the effect of nucleus with the magnetic field. The last one is quadrupole effect of nucleus. It can be added different terms in Eq.13 (Slichter, 1963).

5.2. $\vec{g}$ tensor

The total magnetic moment of ion is $\vec{\mu} = g \mu_B \vec{J}$. $\vec{J}$ is the ratio of total angular momentum to Planck constant. Landé factor g is depend on $\vec{S}, \vec{L}, \vec{J}$. For the base energy level if $\vec{L}$ is zero, g factor is equal free electron's g-factor. But, g-factor in the exited energy levels separated from the g-factor of free electron. The hamiltonian for an ion which is in the magnetic field is following (Weil, 1994).

$$\hat{H} = \mu_B \vec{H} \cdot (\hat{L} + g_e \hat{S}) + \lambda \hat{L} \cdot \hat{S} \tag{14}$$

In this equation, first term is Zeeman effects, second one is spin-orbit interaction. The first order energy of ion which shows $|J, M\rangle$ and it is excepted not degenerate is seen at below.

$$E_J = \langle J, M | g_e \mu_B H_z \hat{S}_z | J, M \rangle + \langle J, M | (\mu_B H_Z + \lambda \hat{S}_z) | J, M \rangle \tag{15}$$

We can write the hamiltonian equation which uses energy equations. There is two terms in the hamiltonian equations. The first term is the independent temperature coefficient for paramagnetic for paramagnetic, the last terms are only for spin variables. If the angular moment of ion occurs because of spin, $\vec{g}$-tensor is to be isotropic.

6. Ferromagnetic resonance

The most important parameters for ferromagnet can be deduced by the ferromagnetic resonance method. FMR absorption curves may be obtained from Eq.(12) by chancing frequency or magnetic field. FMR signal can be detected by the external magnetic field and frequency such as EPR signal. The field derivative FMR absorption spectra are greater than in EPR as a generally. The linear dependence of frequency of resonance field may be calculated from 1 GHz to 100 GHz range in frequency spectra (L-, S-, C-, X-, K-, Q-, V-, E-, W-, F-, and D-band). The resonance frequency, relaxation, linewidth, Landé g-factor (spectroscopic g-factor), the coercive force, the anisotropy field, shape of the specimen, symmetry axes of the crystal and temperature characterized FMR spectra. The broadening of the FMR absorption line depend on the line width (so called $1/T_2$ on the Bloch-Bloembergen type damping form). The nonuniform modes are seen in the EPR signal. The nonlinear effects for FMR are shown by the relationship between the uniform precessions of magnetic moments. The paramagnetic excitation of unstable oscillation of the phonons displays magneto-elastic interaction in ferromagnetic systems. This behaviour so called magnetostriction. The FMR studies have led to the development of many micro-wave devices. These phenomenon are microwave tubes, circulators, oscillators, amplifiers, parametric frequency converters, and limiters. The resonance absorption curve of electromagnetic waves at centimeter scale by ferromagnet was first observed by Arkad'ev in 1913 (Arkad'ev, 1913) .

The sample geometry, relative orientation of the equilibrium magnetization $\vec{M}$, the applied dc magnetic field $\vec{H}$ and experimental coordinate systems are shown in Fig.4.

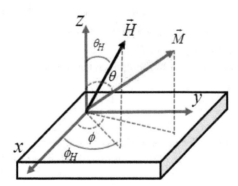

Figure 4. Sample geometries and relative orientations of equilibrium magnetization $\vec{M}$ and the dc components of external magnetic field, $\vec{H}$ for thin films.

The ferromagnetic resonance data analyzed using the free energy expansion similar to that employed

$$E_T = E_Z + E_a + E_d + E_{ex} + ...$$
$$E_Z = -M \cdot H \left(\sin\theta \sin\theta_H \cos(\phi - \phi_H) + \cos\theta \cos\theta_H \right)$$
$$E_a = K_{u1} \sin^2\theta + K_{u2} \sin^4\theta + K_{u3} \sin^6\theta ... \qquad (16)$$
$$E_d = -2\pi M^2 \sin^2\theta$$
$$E_{ex} = -JS_iS_j$$

Where, E_Z, E_a, E_d, E_{ex} are Zeeman, magnetocrystalline anisotropy, demagnetization and ferromagnetic exchange energy. (θ, ϕ) and (θ_H, ϕ_H) are the angles for magnetization and applied magnetic field vector in the spherical coordinates, respectively. Magnetic anisotropy energy arises from either the interaction of electron spin magnetic moments with the lattice via spin-orbit coupling. On the other hand, anisotropy energy induced due to local atomic ordering. The θ in anisotropy energy is the angle between magnetization orientation and local easy axis of the magnetic anisotropy. K_{u1} and K_{u2} are energy density constants. The demagnetization field is proportional to the magnetic free pole density. The exchange energy for thin magnetic film may be neglected in generally. Because associated energies is small. But, this exchange energy are not neglected for multilayer structures. This energy occurs between the magnetic layers, so that this energy called interlayer exchange energy. This expression is seen at the end of this subject in details. $K_{eff} = \pi M^2 + K_U$ is the effective uniaxial anisotropy term and K_u takes into account some additional second-order uniaxial anisotropy and $H_{eff} = 2\pi M_S + (2K_u/M_s)$ is the effective field for a single magnetic films. The equilibrium values of polar angles θ for the magnetization vector $\vec{M}$ are obtained from static equilibrium conditions. $E_\theta, E_\phi, E_{\theta\theta}$ and $E_{\phi\phi}$ can be easily calculated using the Eq. (16). Neglecting the damping term one can write the equation of motion for the magnetization vector $\vec{M}$ as

$$\frac{1}{\gamma}\frac{d\vec{M}}{dt} = \vec{M} \times \vec{H}_{eff}. \qquad (17)$$

Here the $\vec{H}_{eff}$ is the effective magnetic field that includes the applied magnetic field and the internal field due to the anisotropy energy. The dynamic equation of motion for magnetization with the Bloch-Bloembergen type damping term is given as Eq.(18).

$$\frac{1}{\gamma}\frac{d\vec{M}}{dt} = \vec{M} \times \vec{H}_{eff} - \frac{\vec{M} - \delta_{iz}M_0}{T}. \qquad (18)$$

Here, $T = (T_2, T_2, T_1)$ represents both transverse (for M_x and M_y components) and the longitudinal (for M_z components) relaxation times of the magnetization. That is, T_1 is the spin-lattice relaxation time, T_2 is the spin-spin relaxation time, and $\delta_{iz} = (0,0,1)$ for (x,y,z) projections of the magnetization. In the spherical coordinates the Bloch-Bloembergen equation can be written as below;

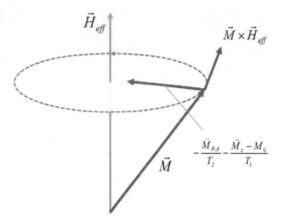

Figure 5. Damped precession of a magnetic moment $\vec{M}$ toward the effective magnetic field $\vec{H}_{eff}$ according to the Bloch-Bloembergen type equation (Aktaş, 1993, 1994; Yalçın, 2008(a)).

$$\frac{1}{\gamma}\frac{d\vec{M}}{dt} = \frac{\vec{M}}{\left|\vec{M}\right|} \times \vec{\nabla}E - \frac{\vec{M}_{\theta,\phi}}{\gamma T_2} - \frac{\vec{M}_z - M_0}{\gamma T_1}. \tag{19}$$

Where, the torque is obtained from the energy density through the expression

$$\vec{\nabla}E = -\left(\frac{\partial E}{\partial \theta}\right)\hat{e}_\phi + \frac{1}{\sin\theta}\left(\frac{\partial E}{\partial \phi}\right)\hat{e}_\theta. \tag{20}$$

For a small deviation from the equilibrium orientation, the magnetization vector $\vec{M}$ can be approximated by

$$\vec{M} = M_s\hat{e}_r + m_\theta\hat{e}_\theta + m_\phi\hat{e}_\phi. \tag{21}$$

Where the dynamic transverse components are assumed to be sufficiently small and can be given as

$$\begin{aligned}
m_\theta(z,t) &= m_\theta^0 \exp i(\omega t \pm kz) \\
m_\phi(z,t) &= m_\phi^0 \exp i(\omega t \pm kz)
\end{aligned} \tag{22}$$

Dispersion relation for films can be derived by using these solutions (Eq.(22)) in Eqs. (19) and (20). On the other hand, the eigen frequency of thin films mode is determined by the static effective field and can be derived directly from the total free energy for magnetic system/ferromagnet. It is given by the second derivatives of the total energy with respect to the θ and ϕ (Smit, 1955; Artman, 1957; Wigen, 1984, 1988, 1992; Baseglia, 1988; Layadi, 1990; Farle, 1998). The matrices form for m_θ and m_ϕ is calculated using the Eq.(19) with Eq.(20, 21, 22).

$$
\begin{pmatrix}
i\dfrac{\omega}{\gamma} + \dfrac{1}{\gamma T_2} + \dfrac{E_{\theta\phi}}{M_s \sin\theta} & \dfrac{E_{\phi\phi}}{M_s \sin^2\theta} \\[4mm]
-\dfrac{E_{\theta\theta}}{M_s} & i\dfrac{\omega}{\gamma} + \dfrac{1}{\gamma T_2} - \dfrac{E_{\theta\phi}}{M_s \sin\theta}
\end{pmatrix}
\cdot
\begin{pmatrix} m_\theta \\ m_\phi \end{pmatrix} = 0
\tag{23a}
$$

$$
\left(\frac{\omega}{\gamma}\right)^2 = \frac{1}{M^2 \sin^2\theta}\left(E_{\theta\theta}E_{\phi\phi} - E_{\theta\phi}^2\right) + \left(\frac{1}{\gamma T_2}\right)^2
\tag{23b}
$$

Here $(\omega/\gamma) = g\mu_B H$ is the Larmour frequency of the magnetization in the external dc effective magnetic field. This dispersion relation can be related as the angular momentum analogue to be linear momentum oscillator described $\omega = \sqrt{\kappa/\mu_i}$. Here, restoring force constant κ is the second derivative of the potential part in the energy of system $\kappa = E_{xx}$. The inverse mass μ_i^{-1} is given by the second derivative of the kinetic part in the energy with respect to linear momentum $\mu_i^{-1} = E_{pp}$. The restoring constant in this chapter corresponds to $E_{\theta\theta}$. The inverse mass is proportional to $E_{\phi\phi}$. The $E_{\theta\phi}$ arises when the coordinate system is not parallel to the symmetry and last term originated from relaxation term in Eq.(23) (Sparks, 1964; Morrish, 1965; Vittoria, 1993; Gurevich, 1996; Chikazumi, 1997)

The power absorption from radio frequency (rf) field in a unit volume of sample is given by

$$
P = \frac{1}{2}\omega\chi_2 h_1^2.
\tag{24}
$$

where ω is the microwave frequency, h_1 is the amplitude of the magnetic field component and χ_2 is the imaginary part of the high-frequency susceptibility. The field derivative FMR absorption spectrum is proportional to $d\chi_2/dH$ and the magnetic susceptibility χ is given as

$$
\chi = 4\pi\left(\frac{m_\phi}{h_\phi}\right)_{\phi=0}
\tag{25}
$$

The theoretical absorption curves are obtained by using the imaginary part of the high frequency magnetic susceptibility as a function of applied field (Öner, 1997; Min, 2006; Cullity, 2009)

$$
\chi = \chi_1 + i\chi_2 = \frac{4\pi M_s\left(\dfrac{E_{\theta\theta}}{M_s}\right)\left(\left(\dfrac{\omega_0}{\gamma}\right)^2 - \left(\dfrac{\omega}{\gamma}\right)^2 + i\left(\dfrac{2\omega}{\gamma^2 T_2}\right)\right)}{\left(\left(\dfrac{\omega_0}{\gamma}\right)^2 - \left(\dfrac{\omega}{\gamma}\right)^2\right)^2 + \left(\dfrac{2\omega}{\gamma^2 T_2}\right)^2}
\tag{26}
$$

The dispersion relation can be derived by substituting Eq.(16) into Eq. (23) (Aktaş, 1997; Yalçın, 2004(a), 2004(b), 2008(a); Güner, 2006; Kharmouche, 2007; Stashkevich, 2009)

$$\left(\frac{\omega}{\gamma}\right)^2 = \left[H\cos(\theta-\theta_H) + H_{eff}\cos(2\theta)\right]\cdot\left[H\cos(\theta-\theta_H)+H_{eff}\cos^2\theta\right]+\left(\frac{1}{\gamma T_2}\right)^2 \qquad (27)$$

here, $\omega_0 = 2\pi\upsilon$ is the circular frequency of the EPR spectrometer. Fitting Eq.(27) with experimental results of the FMR measurement at different out-of-plane-angle (θ_H), the values for the effective magnetization can be obtained.

Figure 6 uses of both experimental and theoretical coordinate systems for the nanowire sample geometry. Equilibrium magnetization $\vec{M}$ and dc-magnetic field $\vec{H}$ are shown in this figure and also the geometric factor and hexagonal nanowire array presentation of nanowire are displayed. The ferromagnetic resonance theory has been developed for thin films applied to nanowires with the help of the following Fig.6. The effective uniaxial anisotropy term for nanowire arrays films $K_{eff} = \pi M^2\left(1-3P\right)+K_U$ is written in this manner for arrayed nanowires. The first term in the K_{eff} is due to the magnetostatic energy of perpendicularly-arrayed NWs (Dubowik, 1996; Encinas-Oropesa, 2001; Demand, 2002; Yalçın, 2004(a); Kartopu, 2009, 2010, 2011(a)) and constant with the symmetry axis along wire direction. The second term in the K_{eff} is packing factor for a perfectly ordered hcp NW arrays. The packing factor is defined as $P=(\pi/2\sqrt{3})(d/r)^2$. The packing factor (P) of nanowires increases, nanowire diameter increases, the preferential orientation of the easy direction of magnetization changes from the parallel to the perpendicular direction to the wire axis (Kartopu, 2011(a)). As further, the effective uniaxial anisotropy (K_{eff}) for a perfectly ordered hcp NWs should decrease linearly with increasing packing factor. $H_{eff} = 2\pi M_S\left(1-3P\right)+\left(2K_u/M_s\right)$, which is the effective anisotropy field derived from the total magnetic anisotropy energy of NWs Eq. (16). The values for total magnetization have been obtained by fitting H_{eff} with experimental results of FMR measurements at different angles (θ_H) of external field $\vec{H}$. The experimental spectra are proportional to the derivative of the absorbed power with respect to the applied field which is also proportional to the imaginary part of the magnetic susceptibility.

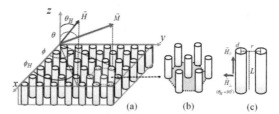

(a)   (b)   (c)

Figure 6. (a) Schematic representation of the cobalt nanowires and the relative orientation of the equilibrium magnetization M and the dc component of the external magnetic field H, for the FMR experiments and their theoretical calculations. (b) Hexagonal NW array exhibiting a total of seven wires and the dashed lines bottom of the seven wires indicate the six fold symmetry. (c) Sample parameters used in the packing factors P calculation.

The experimental data were analyzed by using magnetic energy density for a system consisting of n magnetic layers with saturation magnetization M_s and layer thickness t_i. The magnetic energy density for the nanoscale multilayer structures the energy per unit surface area can be written as below

$$
E = - \left(
\begin{array}{l}
\sum\limits_{i=1}^{n} t_i M_s H \left(\sin\theta_i \sin\theta_H \cos(\phi_i - \phi_H) + \cos\theta_i \cos\theta_H \right) + \\[2mm]
\sum\limits_{i=1}^{n} t_i K_{a,i} \sin^2\phi_i \cos^2\theta_i + \sum\limits_{i=1}^{n} t_i K_{b,i} \sin^4\phi_i \cos^4\theta_i + \\[2mm]
\sum\limits_{i=1}^{n-1} A_{i,i\mp1} \left(\sin\theta_i \sin\theta_{i\mp1} \cos(\phi_{i\mp1} - \phi_i) + \cos\theta_i \cos\theta_{i\mp1} \right) + \\[2mm]
\sum\limits_{i=1}^{n-1} B_{i,i\mp1} \left(\sin\theta_i \sin\theta_{i\mp1} \cos(\phi_{i\mp1} - \phi_i) + \cos\theta_i \cos\theta_{i\mp1} \right)^2
\end{array}
\right)
\tag{28}
$$

Where, θ_i is the polar angle of the magnetization M_s to the z-axis and φ_i is the azimuth angle to the x-axis in the film plane. The first term is the Zeeman energy. The second and third terms correspond to first and second order magnetocrystalline energy with respectively. These energies due to the demagnetization field and any induced perpendicular anisotropy energy. On the other hand, these energies qualitatively have the same angular dependence with respect to the film normal. The second order magnetocrystalline energy term can be neglected for most of the ferromagnetic systems. The last two terms corresponds to bilinear and biquadratic interactions of ferromagnetic layers through nonmagnetic spacer via conduction energies. $A_{i,i\mp1}$ and $B_{i,i\mp1}$ are bilinear and biquadratic coupling constants, respectively. The bilinear exchange interaction can be written from Eq.16. $A_{i,i\mp1}$ can be either negative and positive depending on antiferromagnetic and ferromagnetic interactions, respectively. The antiparallel/perpendicular and parallel alignments of magnetization of nearest neighboring layers are energetically favorable for a negative/positive value of $B_{i,i\mp1}$. Biquadratic interaction for spin systems have been analysed for Ising system in detail (Chen, 1973; Erdem, 2001). The biquadratic term is smaller than the bilinear interaction term. Therefore, it can be neglected for most of the ferromagnetic systems. The indirect exchange energy depends on spacer thickness and even shows oscillatory behavior with spacer thickness (Ruderman, 1954; Yosida, 1957; Parkin, 1990, 1991(a), 19901(b), 1994). The current literature on single ultrathin films and multilayers is given in below at table (Layadi, 1990(a), 1990(b); Wigen, 1992; Zhang, 1994(a), 1994 (b); Goryunov, 1995; Ando, 1997; Farle; 1998; Platow, 1998; Schmool, 1998; Lindner, 2003; Sklyuyev, 2009; Topkaya, 2010; Erkovan, 2011).

This type exchange-coupling system is located in an external magnetic field, the magnetic moment in each layer. The suitable theoretical expression may be derived in order to deduce magnetic parameter for ac susceptibility. The equation of precession motion for magnetization of the ith layer in the spherical coordinates with the Bloch-Bloembergen type relaxation term can be written as

$$
\frac{1}{\gamma}\frac{d\vec{M}}{dt} = \frac{1}{t_i}\frac{\vec{M}}{M_{i,s}} \times \vec{\nabla}_{M_i} E - \frac{\vec{M}_{\theta i,\varphi i}}{\gamma T_2} - \frac{\vec{M}_{z,i} - M_{i,s}}{\gamma T_1}. \quad (i = 1, n)
\tag{29}
$$

The matrices form for $m_{\theta,i-1}$, $m_{\phi,i-1}$, $m_{\theta,i}$, $m_{\phi,i}$, $m_{\theta,i+1}$ and $m_{\phi,i+1}$ of each magnetic layers calculated using the Eq.(29) with Eq.(20, 21,28).

$$
\begin{pmatrix}
\Omega+\dfrac{E_{\theta_1\phi_1}}{t_1M_s\sin\theta_1} & \dfrac{E_{\phi_1\phi_1}}{t_1M_s\sin^2\theta_1} & \dfrac{E_{\theta_1\phi_1}}{t_1M_s\sin\theta_1} & \dfrac{E_{\phi_1\phi_2}}{t_1M_s\sin\theta_1\sin\theta_2} & \dfrac{E_{\theta_1\phi_3}}{t_1M_s\sin\theta_1} & \dfrac{E_{\phi_1\phi_3}}{t_1M_s\sin\theta_1\sin\theta_3} \\[2mm]
\dfrac{E_{\theta_1\theta_1}}{t_1M_s} & \Omega-\dfrac{E_{\theta_1\phi_1}}{t_1M_s\sin\theta_1} & \dfrac{E_{\theta_1\theta_2}}{t_1M_s} & -\dfrac{E_{\theta_1\phi_2}}{t_1M_s\sin\theta_2} & \dfrac{E_{\theta_1\theta_3}}{t_1M_s} & \dfrac{E_{\theta_1\phi_3}}{t_3M_s\sin\theta_3} \\[2mm]
\dfrac{E_{\theta_2\phi_2}}{t_2M_s\sin\theta_2} & \dfrac{E_{\phi_1\phi_2}}{t_2M_s\sin\theta_1\sin\theta_2} & \Omega+\dfrac{E_{\theta_2\phi_2}}{t_2M_s\sin\theta_2} & \dfrac{E_{\phi_2\phi_2}}{t_2M_s\sin^2\theta_2} & \dfrac{E_{\theta_2\phi_3}}{t_2M_s\sin\theta_3} & \dfrac{E_{\phi_2\phi_3}}{t_2M_s\sin\theta_2\sin\theta_3} \\[2mm]
\dfrac{E_{\theta_2\theta_2}}{t_2M_s} & -\dfrac{E_{\theta_2\phi_1}}{t_2M_s\sin\theta_1} & \dfrac{E_{\theta_2\theta_2}}{t_2M_s} & \Omega-\dfrac{E_{\theta_2\phi_2}}{t_2M_s\sin\theta_2} & \dfrac{E_{\theta_2\theta_3}}{t_3M_s} & \dfrac{E_{\theta_2\phi_3}}{t_3M_s\sin\theta_3} \\[2mm]
\dfrac{E_{\theta_3\phi_3}}{t_3M_s\sin\theta_3} & \dfrac{E_{\phi_1\phi_3}}{t_3M_s\sin\theta_1\sin\theta_3} & \dfrac{E_{\theta_2\phi_3}}{t_3M_s\sin\theta_2} & \dfrac{E_{\phi_2\phi_3}}{t_3M_s\sin\theta_2\sin\theta_3} & \Omega+\dfrac{E_{\theta_3\phi_3}}{t_3M_s\sin\theta_3} & \dfrac{E_{\phi_3\phi_3}}{t_3M_s\sin^2\theta_3} \\[2mm]
\dfrac{E_{\theta_3\theta_3}}{t_3M_s} & -\dfrac{E_{\theta_3\phi_1}}{t_3M_s\sin\theta_3} & \dfrac{E_{\theta_2\theta_3}}{t_3M_s} & -\dfrac{E_{\theta_3\phi_2}}{t_3M_s\sin\theta_2} & \dfrac{E_{\theta_3\theta_3}}{t_3M_s} & \Omega-\dfrac{E_{\theta_3\phi_3}}{t_3M_s\sin\theta_3}
\end{pmatrix}
\cdot
\begin{pmatrix} m_{\theta_1} \\ m_{\varphi_1} \\ m_{\theta_2} \\ m_{\varphi_2} \\ m_{\theta_3} \\ m_{\varphi_3} \end{pmatrix} = 0
\tag{30}
$$

1, 2, 3 number in this Eq.(30) corresponds to $i-1$, i and $i+1$, respectively. Here $\Omega = i\left(\dfrac{\omega}{\gamma}\right)+\dfrac{1}{\gamma T_2}$. Then dispersion relation for ferromagnetic exchange-coupled n-layers has been calculated using the (2nx2n) matrix on the left-hand side of Eq. (30) in below in detail.

$$
\left(\frac{\omega}{\gamma}\right)^{2n}+C_{(2n-2)/2}\left(\frac{\omega}{\gamma}\right)^{2n-2}+...+C_1\left(\frac{\omega}{\gamma}\right)^2+C_0=0
\tag{31}
$$

Here, n is the number of ferromagnetic layer. C_0, C_1,... etc. are constant related to t_i, M_s, $E_{\theta i\theta i}$, $E_{\varphi i\varphi i}$, $E_{\theta i\varphi i}$, $\sin\theta_i$ and $\sin^2\theta_i$. The dispersion relations for monolayer, trilayers, five-layers obtained from the Eq.(31). For tri-layers detail information are seen in ref. (Zhang, 1994(a); Schmool, 1998; Lindner, 2003). It is given that the dispersion relation for monolayer, trilayers, five-layers and multilayers/n-layers in Fig. 7.

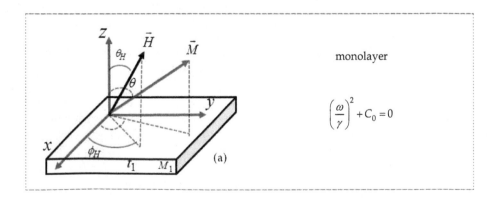

monolayer

$$\left(\frac{\omega}{\gamma}\right)^2+C_0=0$$

(a)

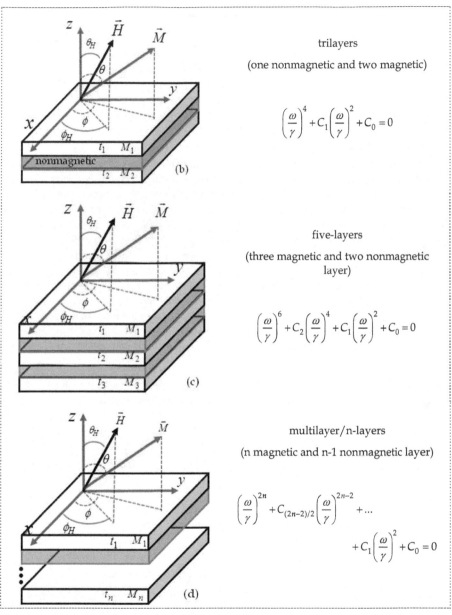

trilayers

(one nonmagnetic and two magnetic)

$$\left(\frac{\omega}{\gamma}\right)^4 + C_1\left(\frac{\omega}{\gamma}\right)^2 + C_0 = 0$$

(b)

five-layers

(three magnetic and two nonmagnetic layer)

$$\left(\frac{\omega}{\gamma}\right)^6 + C_2\left(\frac{\omega}{\gamma}\right)^4 + C_1\left(\frac{\omega}{\gamma}\right)^2 + C_0 = 0$$

(c)

multilayer/n-layers

(n magnetic and n-1 nonmagnetic layer)

$$\left(\frac{\omega}{\gamma}\right)^{2n} + C_{(2n-2)/2}\left(\frac{\omega}{\gamma}\right)^{2n-2} + \ldots$$

$$+ C_1\left(\frac{\omega}{\gamma}\right)^2 + C_0 = 0$$

(d)

Figure 7. Schematic representation of the (a) one layer, (b) three layer, (c) five layer and (d) n magnetic layer and their relative orientation of the equilibrium magnetization $\vec{M}$ and the dc component of the external magnetic field $\vec{H}$ for the FMR experiments and their theoretical calculations.

6. Example: Exchange spring (hard/soft) behaviour

The Bloch wall, Néel line and magnetization vortex are well known properties for magnetic domain in magnetic systems. The multilayer structures are ordered layer by layer. The best of the sample for multilayer structure are exchange-spring systems. The equilibrium magnetic properties of nano-structured exchange-spring magnets may be studied in detail for some selected magnetic systems. The exchange systems are oriented from the exchange coupling between ferromagnetic and antiferromagnetic films or between two ferromagnetic films. This type structure has been extensively studied since the phenomenon was discovered (Meiklejohn, 1956, 1957). Kneller and Hawing have been used firstly the "exchange-spring" expression (Kneller, 1991). Spring magnet films consist of hard and soft layers that are coupled at the interfaces due to strong exchange coupling between relatively soft and hard layers. The soft magnet provides a high magnetic saturation, whereas the magnetically hard material provides a high coercive field. Skomski and Coey explored the theory of exchanged coupled films and predicted that a huge energy about three times of commercially available permanent magnets (120 MGOe) can be induced (Skomski, 1993; Coey, 1997). The magnetic reversal proceeds via a twisting of the magnetization only in the soft layer after saturating hard layers, if a reverse magnetic field that is higher than exchange field is applied. The spins are sufficiently closed to the interface are pinned by the hard layer, while those in deep region of soft layer rotate up to some extent to follow the applied field (Szlaferek, 2004). To be more specific, the angle of the rotation depends on the distance to the hard layer. That is the angle of rotating in a spiral spin structure similar to that of a Bloch domain wall. If the applied field is removed, the soft spins rotate back into alignment with the hard layer.

The general expression of the free energy for exchange interaction spring materials at film ($\theta_{i,i\pm1} = \pi/2$ and $\theta_H = \pi/2$) plane in spherical coordinate system as below.

$$
\begin{aligned}
E = & -\sum_{i=1}^{N} \vec{H} \cdot \vec{M}_i - \sum_{i=1}^{N} K_{a,i} \cos^2 \phi_i - \sum_{i=1}^{N} K_{b,i} \cos^4 \phi_i \\
& - \sum_{i=1}^{N-1} \frac{A_{i,i+1}}{t^2} \cos\left(\phi_{i\mp1} - \phi_i\right) - \sum_{i=1}^{N-1} \frac{B_{i,i+1}}{t^2} \cos\left(\phi_{i\mp1} - \phi_i\right)^2
\end{aligned}
\tag{32}
$$

The expression is obtained as following using $\phi_i \to \phi_i'$ for magnetization's equilibrium orientations of each layer at a state of equilibrium under the external magnetic field.

$$
\tan \phi_i' =
$$
$$
\frac{HM_i t^2 \sin\phi_H + A_{i,i\mp1} \sin\phi_{i\mp1} + 2B_{i,i\mp1} \sin\phi_{i\mp1} \cos(\phi_i - \phi_{i\mp1})}{HM_i t^2 \cos\phi_H + 2t^2 K_{a,i} \cos\phi_i + 4t^2 K_{b,i} \cos^3 \phi_i + A_{i,i\mp1} \cos\phi_{i\mp1} + 2B_{i,i\mp1} \cos\phi_{i\mp1} \cos(\phi_i - \phi_{i\mp1})}
\tag{33}
$$

In this example, second-order anisotropy term ($K_{b,i} = 0$) and biquadratic interaction constant ($B_{i,i\pm1} = 0$) considered and the result obtained show as following as adapted with spring magnets SmCo(hard)/Fe(soft). For theoretical analysis, the exchange-spring magnet SmCo/Fe is divided into subatomic multi-layers (d=2 Å), and the spins in each layer are characterized by the average magnetization M_i, and the uniaxial anisotropy constant K_i, (Fig.8).

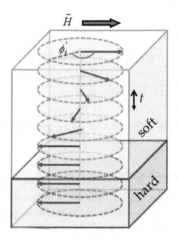

Figure 8. Schematic illustration of phases of exchange spring magnets.

Sublayers are coupled by an exchange constant $A_{i,i+1}$ (Astalos, 1998; Fullerton, 1998, 1999; Jiang, 1999, 2002, 2005; Grimsditch, 1999; Scholz, 2000; Hellwig, 2000; Pollmann, 2001; Dumesnil, 2002). ϕ_i is the angle formed by the magnetization of the i th plane with the in-plane (where the external field is always perpendicular to the film normal) easy axis of the hard layer (Yıldız, 2004(a), 2004(b)). The FMR spectra for exchange-spring magnet of SmCo/Fe have been analyzed using the Eqs. (26, 27, and 33) in Fig.9. Sm-Co (200 Å)/Fe (200 Å and 100 Å) bilayers have been grown on epitaxial 200 Å Cr(211) buffer layer on single crystal MgO(110) substrates by magnetron sputtering technique (Wüchner, 1997). To prevent oxidation Sm-Co/Fe film was coated with a 100 Å thick Cr layer. The FMR spectra for exchange-spring magnets of 200 Å and 100 Å Fe samples for different angles of the applied magnetic field in the film plane are presented in Fig.9.

There are three peaks that are one of them corresponds to the bulk mode and the remaining to the surface modes for 200 Å Fe sample. For more information about the FMR studies exchange spring magnets look at the ref. (Yildiz, 2004(a), 2004(b)). Exchange-spring coupled magnets are promising systems for applications in perpendicular magnetic data recording-storage devices and permanent magnet (Schrefl, 1993(a), 1993(b),1998, 2002; Mibu, 1997, 1998).

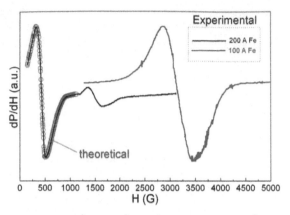

Figure 9. FMR spectra for SmCo(200 Å)/Fe(200 Å) (black line) and SmCo(200 Å)/Fe(100 Å) (blue line) samples. These FMR spectra originated from the iron/soft layers.

7. Superparamagnetic resonance

Magnetic nanoparticles have been steadily interested in science and nanotechnology. As the dimensions of magnetic nanoparticles decrease to the nanometer scale, these nanoparticles start to exhibit new and interesting physical properties mainly due to quantum size effects (Yalçın, 2004(a), 2008(b), 2012). A single domain particle is commonly referred to as superparamagnetic (Held, 2001; Diaz, 2002; Fonseca, 2002). The superparamagnetic/single-domain nanoparticles are important for non surgical interfere of human body. Even the intrinsic physical characteristics of nanoparticles are observed to change drastically compared to their macroscopic counterparts. Stoner-Wohlfarth (Stoner, 1948) and Heisenberg model (Heisenberg, 1928) to describe the fine structure were firstly used in detail. A simple (Bakuzis, 2004) and the first atomic-scale models of the ferrimagnetic and heterogeneous systems in which the exchange energy plays a central role in determining the magnetization of the NPs, were studied (Kodama, 1996, 1999; Kodama & Berkowitz, 1999). Superparamagnetic resonance (SPR) studies of fine magnetic nanoparticles is calculated a correlation between the line-width and the resonance field for superparamagnetic structures (Berger, 1997, 1998, 2000(a), 2000(b), 2001; Kliava, 1999). The correlation of the line-width and the resonance field is calculated from Bloch-Bloembergen equation of motion for magnetization. The SPR spectra, line width and resonance field may be analyzed by using the Eq.(34) in below. The equation of motion for magnetization with Bloch-Bloembergen type relaxation term for FMR adapted for superparamagnetic structures from Eqs.(18) and (19) in below.

$$\chi_2(H) = \frac{1}{\pi} \frac{\Delta_H(\Delta_H^2 + H^2 + H_r^2)}{\left(\Delta_H^2 + (H - H_r)^2\right) \cdot \left(\Delta_H^2 + (H + H_r)^2\right)} \tag{34}$$

Here, $\Delta_H = 1/\gamma T_2$, $H_r = -(\omega/\gamma)$. This equation for SPR system so called modified Bloch for fine particle magnets. The SPR microwave absorption is proportional to the imaginary part of the dynamic susceptibility. The line shape and resonance field for superparamagnet is obtained. The temperature evolution for the SPR line-width for nanoparticles can be calculated by $\Delta_H = \Delta_T L(x)$. In this expression Δ_T is a saturation line-width at a temperature T, $L(x) = \coth(x) - (1/x)$ is the Langevin function with $x = MVH_{eff}/k_B T$, V is the particle volume. The superparamagnetic (Chastellain, 2004; Dormer, 2005; Hamoudeh, 2007), core-shell nanoparticles and nanocrystalline nanoparticles (Woods, 2001; Wiekhorst, 2003; Tartaj, 2004) have been performed for possible biological applications (Sun, 2005; Zhang, 2008). In additional, superparamagnetic nanoparticles have been used for hydrogels, memory effects and electronic devices (Raikher, 2003; Sasaki, 2005; Heim, 2007).

8. Result and discussions

The EPR, FMR and SPR signals have been observed in Fig.10. The EPR signal has reached approaching peak level about 3000 G as seeing at Fig.10. It's symmetric and line width are narrower than resonance field, in generally. If EPR samples show crystallization, resonance field value starts to change. The EPR signal can be observed at lower temperature about 3000 G and the signal can show crystalline property. The signal is observed in two different areas at FMR spectra as the magnetic field is parallel and perpendicular to the film. The FMR spectra are observed at low field when the magnetic field is parallel to the film, in generally. On the other hand, the FMR spectra are observed at highest field when the magnetic field is perpendicular to the film. For other conditions FMR signals are observed between these two conditions for thin films. FMR spectra can be seen a wide range of field so as to the thin films are full.

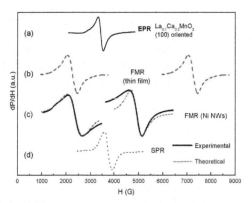

Figure 10. (a) The EPR/ESR experimental signal for La$_{0.7}$Ca$_{0.3}$MnO$_3$ samples at room temperature (see, Kartopu, 2011 (b)). (b) Theoretical FMR spectra calculated from Eq. (26) with Eq.(27) at parallel ($\theta = 90°$;~ 2000 G) and perpendicular ($\theta = 0°$;~7000 G) position of OPG case. (c) The theoretical (red-dot line in online) and experimental FMR spectra for Ni NWs (P= 29,6; L=0,8 μm, τ=13) (see for detail, Kartopu, 2011 (a)). (d) The theoretical SPR signal for superparamagnet by Eq.(34) at room temperature.

FMR spectra are similar to thin films at nanowire samples. In case of occupancy rate is that as the theoretical $P < 33\%$ for Nickel (Ni) it behaviors like thin film. But, in case of occupancy rate is that as the theoretical $P < 33\%$ it behaviors different from thin film. This situation is clearly visible from $H_{eff} = 2\pi M_S (1 - 3P) + (2K_u / M_s)$. If the occupancy rate is $P < 33\%$ sample's signals show the opposite behavior according to thin film FMR signals. Look at for more information (Kartopu, 2011 (a)). This is perceived as changes the direction of the easy axis. The changes of easy axes depend on magnetization (Terry, 1917) and porosity (Kartopu, 2011 (a)) for magnetic materials/transition elements. The SPR signal is similar to EPR signal. SPR peak may show symmetrical properties both at room temperature and low temperatures. The SPR signal is in the form of Lorentzian and Gaussian line shapes at all temperature range. Specially prepared nanoparticles SPR peak exhibit shift in symmetry. The line width of SPR peak expands at low temperature.

Author details

Orhan Yalçın
Niğde University, Niğde, Turkey

Acknowledgement

I would like to thank **Muhittin Öztürk** and **Songül Özüm** of Niğde University for valuable discussions an the critical reading of the chapter. This study was supported by Research found (Grant No. FEB2012/12) of Niğde University.

9. References

Adeyeye, A.O.; Bland, J.A.C.; Daboo, C.; Hasko, D.G. (1997). Magnetostatic interactions and magnetization reversal in ferromagnetic wires. *Phys. Rev. B*. Vol. 56, pp. 3265.

Aktaş, B. (1993). Clear evidence for field induced unidirectional exchange surface anisotropy in NiMn alloys. *Solid State Commun*.Vol. 87, pp. 1067.

Aktaş, B.; Özdemir, M. (1994). Simulated spin wave resonance absorption curves for ferromagnetic thin films and application to NiMn films. *Physica B*. Vol. 119, pp. 125.

Aktaş, B. (1997). FMR properties of epitaxial Fe3O4 films on MgO(100). *Thin Solid Films*. Vol. 307, pp. 250.

Aktaş, B.; Özdemir, M.; Yilgin, R.; Öner, Y.; Sato, T.; Ando, T. (2001). Thickness and temperature dependence of magnetic anisotropies of Ni77 Mn23 films. *Physica B*. Vol. 305, pp. 298.

Aktaş, B.; Yildiz, F.; Rameev, B.; Khaibullin, R.; Tagirov, L.; Özdemir, M. (2004). Giant room temperature ferromagnetism in rutile TiO2 implanted by Co. *Phys. stat. sol. (c)*. Vol. 12, pp. 3319.

Aktaş, B.; Tagirov, L. & Mikailov, F. (October, 2006). *Magnetic Nanostructures*, Springer Series in materials science, Vol. 94, ISBN 978-3-540-49334-1.

Ament, W. S.; Rado, G. T. (1955). Electromagnetic effects of spin wave resonance in ferromagnetic metals. *Phys. Rev.* Vol. 97, pp. 1558.

Ando, Y.; Koizumi, H.; Miyazaki, T. (1997). Exchange coupling energy determined by ferromagnetic resonance in 80 Ni-Fe/Cu multilayer films. *J. Magn. Magn. Mater.* Vol. 166, pp. 75.

Anisimov, A. N.; Farle, M.; Poulopoulos, P.; Platow, W.; Baberschke, K.; Isberg, P.; Wäppling, R.; Niklasson, A. M. N.; Eriksson, O. (1999). Orbital magnetism and magnetic anisotropy probed with ferromagnetic resonance. *Phys. Rev. Lett.* Vol. 82, pp. 2390.

An, S. Y.; Krivosik, P.; Kraemer, M. A.; Olson, H. M.; Nazarov, A. V.; Patton, C. E. (2004). High power ferromagnetic resonance and spin wave instability processes in permalloy thin films. *J. Appl. Phys.* Vol. 96, pp. 1572.

Anderson, P. W. (1953). Exchange narrowing in paramagnetic resonance. *Rev. Mod. Phy.* Vol. 25, pp. 269.

Arias, R.; Mills, D. L. (1999). Extrinsic contributions to the ferromagnetic resonance response in ultrathin films, *Phys. Rev. B.* Vol. 60, pp. 7395.

Arias, R.; Mills, D. L. (2003). Theory of collective spin waves and microwave response of ferromagnetic nanowire arrays. *Phys. Rev. B.* Vol. 67, pp. 094423.

Arkad'ev, V.K. (1913). The Reflection of Electric Waves from a Wire, *Sov. Phys.-JETP*, Vol. 45A, issue 45, pp. 312.

Artman, J. O. (1957). Ferromagnetic resonance in metal single crystals. *Phys. Rev.* Vol. 105, pp. 74.

Artman, J. O. (1979). Domain mode FMR in materials with K_1 and K_u. *J. Appl. Phys.* Vol. 50, pp. 2024.

Astalos, R. J.; Camley, R. E. (1998). Magnetic permeability for exchange-spring magnets: application to Fe/Sm-Co. *Phys. Rev. B.* Vol. 58, pp. 8646.

Aswal, D.K.; Singh, A.; Kadam, R.M.; Bhide, M.K.; Page, A.G.; Bhattacharya, S.; Gupta, S.K.; Yakhmi, J.V.; Sahni, V.C. (2005). Ferromagnetic resonance studies of nanocrystalline $La_{0.6}Pb_{0.4}MnO_3$ thin films. *Mater. Lett.* Vol. 59, pp. 728.

Azzerboni, B.; Asti, G.; Pareti, L.; Ghidini, M. (2006). Magnetic nanostructures in modern technology, spintronics, magnetic MEMS and recording. Proceedings of the NATO advanced study institute on magnetic nanostructures for micro-electromechanical systems and spintronic applications catona, *Published by Springer*. Italy. ISBN 978-1-4020-6337-4.

Backes, D.; Bedau, D., Liu, H., Langer, J.; Kent, A.D. (2012). Characterization of interlayer interactions in magnetic random access memory layer stacks using ferromagnetic resonance. *J. Appl. Phys.* Vol. 111, pp. 07C721.

Baek, J.S.; Min, S.G.; Yu, S.C.; Lim, W.Y. (2002). Ferromagnetic resonance of Fe–Sm–O thin films. *J. App. Phys.* Vol. 93, pp. 7604.

Bai, Y .; Xu, F.; Qiao, L. (2012). The twice ferromagnetic resonance in hexagonal ferrite single rod and paired rods. *Phys. Lett. A.* Vol.376, pp. 563.

Bakker, F.L.; Flipse, J.; Slachter, A.; Wagenaar, D.; van Wees, B.J. (2012). Thermoelectric detection of ferromagnetic resonance of a nanoscale ferromagnet. *Phys. Rev. Lett.* Vol. 108, pp. 167602.

Bakuzis, A.F.; Morais, P.C. (2004). Magnetic nanoparticle systems: an Ising model approximation. *J. Magn. Magn. Mater.* Vol. 272-276, pp. e1161.

Baseglia, L.; Warden, M.; Waldner, F.; Hutton, S.L.; Drumheller, J.E.; He, Y.Q.; Wigen, P.E.; Maryško M. (1988). Derivation of the resonance frequency from the free energy of ferromagnets. *Phys. Rev. B.* Vol. 38, pp. 2237.

Beguhn, S.; Zhou, Z.; Rand, S.; Yang, X.; Lou, J.; Sun, N.X. (2012). A new highly sensitive broadband ferromagnetic resonance measurement system with lock-in detection. *J. Appl. Phys.* Vol. 111, pp. 07A503.

Berger, R.; Bissey, J.; Kliava, J. (2000(a)). Lineshapes in magnetic resonance spectra. *J. Phys.: Condens. Matter.* Vol. 12, pp. 9347.

Berger, R.; Kliava, J.; Bissey, J.C. (2000(b)). Magnetic resonance of superparamagnetic iron-containing nanoparticles in annealed glass. *J. Appl. Phys.* Vol. 87, pp. 7389.

Berger, R.; Bissey, J.C.; Kliava, J.; Soulard, B. (1997). Superparamagnetic resonance of ferric ions in devitrifield borate glass. *J. Magn. Magn. Mater.* Vol. 167, pp. 129.

Berger, R.; Bissey, J. C.; Kliava, J.; Daubric, H.; Estournès, C. (2001). Temperature dependence of superparamagnetic resonance of iron oxide nanoparticles. *J. Magn. Magn. Mater.* Vol. 234, pp. 535.

Berger, R.; Kliava, J.; Bissey, J. C.; Baïettoz, V. (1998). Superparamagnetic resonance of annealed iron containing borate glass. *J. Phys.: Condens. Matter.* Vol. 10, pp. 8559.

Birkhäuser Verlag, A.G. (2007). Spin glasses statics and dynamics. *Basel, Boston,* Berlin. ISBN 978-3-7643-8999-4.

Birlikseven, C.; Topacli, C.; Durusoy, H.Z.; Tagirov, L.R.; Koymen, A.R.; Aktaş, B. (1999 (a)). Magnetoresistance, magnetization and FMR study of Fe/Ag/Co multilayer film. *J. Magn. Magn. Mater.* Vol. 192, pp. 258.

Birlikseven, C.; Topacli, C.; Durusoy, H.Z.; Tagirov, L.R.; Koymen, A.R.; Aktaş, B. (1999 (b)). Layer-sensitive magnetization, magnetoresistance and ferromagnetic resonance (FMR) study of NiFe/Ag/CoNi trilayer film. *J. Magn. Magn. Mater.* Vol. 202, pp. 342.

Bhagat, S.M.; Anderson, J.R.; Wu, N. (1967). Influence of the anomalous skin effect on the ferromagnetic resonance linewidth in iron. *Phys. Rev.* Vol. 155, pp. 510.

Bhagat, S.M.; Lubitz, P. (1974). Temperature variation of ferromagnetic relaxation in the 3d transition metals. *Phys. Rev. B.* Vol. 10, pp. 179.

Bland, J.A.C.; Heinrich, B. (1994). Ultrathin magnetic structures I: An introduction to the electronic magnetic and structural properties. *Springer-Verlag Berlin Heidelberg.* ISBN 3-540-57407-7.

Bloembergen, B. (1950). On the ferromagnetic resonance in nickel and supermalloy. *Phys. Rev.* Vol. 78, pp. 572.

Bloembergen, N.; Wang, S. (1954). Relaxation effects in para- and ferromagnetic resonace. *Phys. Rev.* Vol. 93, pp. 72.

Borchers, J.A.; Dura, J.A.; Unguris, J.; Tulchinsky, D.; Kelley, M.H.; Majkrzak, C.F. (1998). Observation of antiparallel magnetic order in weakly coupled CoyCu multilayers. *Phys. Rev. Lett.* Vol. 82, pp. 2796.

Bose, T.; Trimper, S. (2012). Nonlocal feedback in ferromagnetic resonance. *Arxiv: 1204-5342. Cond-mat. Mes-hall.* pp. 1.

Boulle, O.; Malinowski, G.; Kläui, M. (2011). Current-induced domain wall motion in nanoscale ferromagnetic elements. *Materials Science and Engineering* R. Vol. 72, pp. 159.

Brown, F.M. (1962).Magnetostatic principles in ferromagnetism. *North Holland Publishing Company.*

Brustolon, M.; Giamello, E. (2009). Electron paramagnetic resonance. *John Wiley & Sons, Inc.* ISBN 978-0-470-25882.

Budak, S.; Yildiz, F.; Özdemir, M.; Aktaş, B. (2003). Electron spin resonance studies on single crystalline Fe_3O_4 films. *J. Magn. Magn. Mater.* Vol. 258–259, pp. 423.

Buschow, K.H.J.; de Boer, F.R. (2004). Physics of magnetism and magnetic materials. *Kluwer Academic Publishers.* ISBN: 0-306-47421-2.

Callen, H.B. (1958). A ferromagnetic dynamical equation. *J. Phys. Chem. Solids.* Vol. 4, pp. 256.

Can, M.M.; Coşkun, M.; Fırat, T. (2012). A comparative study of nanosized iron oxide particles; magnetite (Fe_3O_4), maghemite (γ-Fe_2O_3) and hematite (α-Fe_2O_3), using ferromagnetic resonance. *J. Alloy. Compd.* Vol. 542, pp. 241.

Celinski, Z.; Urquhart, B.; Heinrich, B. (1997). Using ferromagmetic resonance to measure the magnetic moments of ultrathin films, *J. Magn. Magn. Mater.* Vol. 166, pp. 6.

Celinski, Z.; Heinrich, B. (1991). Ferromagnetic resonance line width of Fe ultrathin films grown on a bcc substrate. *J. Appl. Phys.* Vol. 70, pp.5936.

Chappert, C.; Le Dang, K.; Beauvillain, P.; Hurdequint, H.; Renard, D. (1986). Ferromagnetic resonance studies of very thin cobalt films on a gold substrate. *Phys. Rev. B.* Vol. 34, pp. 3192.

Chastellain, M.; Petri, A.; Hofmann, H. (2004). Particle size investigations of a multi-step synthesis of PVA coated superparamagnetic nanoparticles. *J. Colloid. İnterf. Sci.* Vol. 278, pp. 353.

Chen, Y.S.; Cheng, C.W.; Chern, G.; Wu, W.F.; Lin, J.G. (2012). Ferromagnetic resonance probed annealing effects on magnetic anisotropy of perpendicular CoFeB/MgO bilayer. *J. Appl. Phys.* Vol. 111, pp.07C101.

Chen, S.H.; Chang, C.R.; Xiao, J.Q.; Nikolić, K.B. (2009). Spin and charge pumping in magnetic tunnel junctions with precessing magnetization: A nonequilibrium green function approach. *Phys. Rev. B.* Vol. 79, pp. 054424.

Chen, H.H.; Levy, P.M. (1973). Dipole and quadrupole phase transitions in spin-1 models. *Phys. Rev. B. Vol. 7,* 4267.

Chikazumi, S. (1997). Physics of ferromagnetism. *Oxford University Press.* ISBN 0-19-851776-9.

Cochran, J.F. (1995). Light scattering from ultrathin magnetic layers and bilayers in magnetic ultrathin films. Heinrich, B.; Bland, J.A.C. (Eds.) *Springer, Berlin, Heidelberg* Vol. II, pp. 222. B. Hillebrands: Brillouin light scattering in magnetic superlattices. ibid pp. 258.

Cochran, J.F.; Kambersky, V. (2006). Ferromagnetic resonance in very thin films. *J. Magn. Magn. Mater.* Vol. 302, pp. 348.

Coey, J.M.D. (1997). Permanent magnetism. *Solid State Comm.* Vol. 102, pp. 101.

Coey, J.M.D. (2009). Magnetism and magnetic materials. *Cambridge University Press.* ISBN-13 978052181614-4.

Cofield, M.L.; Glocker, D.; Gau, J.S. (1987). Spin-wave resonance in CoCr magnetic thin films. *J. Appl. Phys.* Vol. 61, pp. 3810.

Coşkun, R.; Okutan, M.; Yalçın, O.; Kösemen, A. (2012). Electric and magnetic properties of hydrogels doped with Cu ions. Acta. Phys. Pol. A. Vol. 122, pp. 683.

Crittenden, E.C.Jr; Hoffman, R.W. (1953). Thin films of ferromagnetic materials. *Rev. Mod. Phys.* Vol. 25, pp. 310.

Cullity, B.D.; Graham, C.D. (1990). Introduction to Magnetic Materials. *Wiley.* New York. pp. 199.

Cullity, B.D.; Graham, C.D. (2009). *Introduction to magnetic materials* 2nd Edition.

Damon, R.W. (1953). Relaxation effect in the ferromagnetic resonance. *Rev. Mod. Phys.* Vol. 25, pp. 239.

Damon, R.W. (1963). Ferromagneticre sonance at high power in Rado, G.T.; Suhl, H. (Eds.). *Magnetism.* Vol. I

da Silva, E.C.; Meckenstock, R.; von Geisau, O.; Kordecki, R.; Pelzl, J.; Wolf, J.A.; Grünberg, P. (1993). Ferromagnetic resonance investigations of anisotropy fields of Fe(001) epitaxial layers. *J. Magn. Magn. Mater.* Vol. 121, pp. 528.

Davis, J.A. (1965). Effect of surface pinning on the magnetization of thin films. *J. Appl. Phys.* Vol. 36, pp. 3520.

Demokritov, S.; Rücker, U.; Grünberg, P. (1996). Enhancement of the Curie temperature of epitaxial EuS(100) films caused by growth dislocations. *J. Magn. Magn. Mater.* Vol. 163, pp. 21.

Demokritov, S.; Rücker, U.; Arons, R.R.; Grünberg, P. (1997). Antiferromagnetic interlayer coupling in epitaxial Fe/EuS (100) bilayers. J. Appl. Phys. Vol. 81, pp. 5348.

De Wames, R.E.; Wolfram, T. (1970). Dipole-exchange spinwaves in ferromagnetic films. *J. Appl. Phys.* Vol. 41, pp. 87.

Demand, M.; Encinas-Oropesa, A.; Kenane, S.; Ebels, U.; Huynen, I.; Pirax, L. (2002). Ferromagnetic resonance studies of nickel and permalloy nanowire arrays. *J. Magn. Magn. Mater.*Vol. 249, pp. 228.

De Biasi, E.; Lima, Jr. E.; Ramos, C.A.; Butera, A.; Zysler, R.D. (2013). Effect of thermal fluctuations in FMR experiments in uniaxial magnetic nanoparticles: Blocked vs. superparamagnetic regimes. *J. Magn. Magn. Mater.* Vol. 326, pp. 138.

De Biasi, R.S.; Gondim, E.C. (2006). Use of ferromagnetic resonance to determine the size distribution of γ-Fe$_2$O$_3$ nanoparticles. Solid State Commun. Vol. 138, pp. 271.

De Cos, D.; Arribas, A.G.; Barandiaran, J.M. (2006). Ferromagnetic resonance in gigahertz magneto-impedance of multilayer systems. *J. Magn. Magn. Mater.* Vol. 304, pp. 218.

Diaz, L.L.; Torres, L.; Moro, E. (2002). Transition from ferromagnetism to superparamagnetism on the nanosecond time scale. *Phys. Rev. B.* Vol. 65, pp. 224406.

Dillon, J.F.Jr.; Gyorgy, E.M.; Rupp, L.W.Jr.; Yafet, Y.; Testardi,L.R. (1981). Ferromagnetic resonance in compositionally modulated CuNi films. *J. Appl. Phys.* Vol. 52, pp. 2256.

Ding, J.; Kostylev, M.; Adeyeye, A.O. (2012). Broadband ferromagnetic resonance spectroscopy of permalloy triangular nanorings. *J. Appl. Phys.* Vol. 100, pp. 062401.

Dormer, K.; Seeney, C.; Lewelling, K.; Lian, G.; Gibson, D.; Johnson, M. (2005). Epithelial internalization of superparamagnetic nanoparticles and response to external magnetic field. *Biomaterials*. Vol. 26, pp. 2061.

Dreher, L.; Weiler, M.; Pernpeintner, M.; Huebl, H.; Gross, R.; Brandt, M.S.; Goennenwein, S.T.B. (2012). Surface acoustic wave-driven ferromagnetic resonance in nickel thin films: Theory and experiment. *Arxiv:Cond-mat. Mes-hall*. Vol. 1208, pp. 1.

Dubowik, J. (1996). Shape anisotropy of magnetic heterostructures. *Phys. Rev. B*. Vol. 54, pp.1088.

Dumesnil, K.; Dufour, C.; Mangin, Ph.; Rogalev, A. (2002). Magnetic springs in exchange-coupled D_yFe_2/YFe_2 superlattices: An element-selective x-ray magnetic circular dichroism study. *Phys. Rev. B*. Vol. 65, pp. 094401.

Duraia, E.M.; Abdullin, Kh.A. (2009). Ferromagnetic resonance of cobalt nanoparticles used as a catalyst for the carbon nanotubes synthesis. *J. Magn. Magn. Mater*. Vol. 321, pp. 69.

Durusoy, H.Z.; Aktaş, B.; Yilgin, R.; Terada, N.; Ichikawa, M.; Kaneda, T.; Tagirov, L.R. (2000). New technique for measuring the microwave penetration depth in high- T_c superconducting thin films. *Phys. B*. Vol. 284-288, pp. 953.

Encinas-Oropesa, A.; Demand, M.; Piraux, L.; Huynen, I.; Ebels, U. (2001). Dipolar interactions in arrays of nickel nanowires studied by ferromagnetic resonance. *Phys. Rev. B*. Vol. 63, pp. 104415.

Erdem, R.; Keskin, M. (2001). Dynamics of a spin-1 Ising system in the neighborhood of equilibrium states, *Phys. Rev. E*. Vol. 64, pp. 026102.

Erkovan, M.; Öztürk, S.T.; Topkaya, R.; Özdemir, M.; Aktaş, B.; Öztürk, O. (2011). Ferromagnetic resonance investigation of Py/Cr multilayer system. *J. Appl. Phys*. Vol. 110, pp. 023908.

Estévez, D.C.; Betancourt, I.; Montiel, H. (2012). Magnetization dynamics and ferromagnetic resonance behavior of melt spun FeBSiGe amorphous alloys. *J. Appl. Phys*. Vol. 112, pp. 053923.

Fan, W.J.; Qiu, X.P.; Shi, Z.; Zhou, S.M.; Cheng, Z.H. (2010). Correlation between isotropic ferromagnetic resonance field shift and rotatable anisotropy in polycrystalline NiFe/FeMn bilayers. *Thin Solid Films*. Vol. 518, pp. 2175.

Farle, M. (1998). Ferromagnetic resonance of ultrathin metallic layers. *Rep. Prog. Phys*. Vol. 61, pp. 755.

Farle, M.; Lindner, J.; Baberschke, K. (2000). Ferromagnetic resonance of Ni(111) on Re(0001). *J. Magn. Magn. Mater*. Vol. 212, pp. 301.

Fermin, J.R.; Azevedo, A.; Aguiar, F.M.; Li, B.; Rezende, S.M. (1999). Ferromagnetic resonance linewidth and anisotropy dispersion in thin Fe films. *J. Appl. Phys*. Vol. 85, pp. 7316.

Fischer, H.; Mastrogiacomo, G.; Löffler, J.F.; Warthmann, R.J.; Weidler, P.G.; Gehring, A.U. (2008). Ferromagnetic resonance and magnetic characteristics of intact magnetosome chains in Magnetospirillum gryphiswaldense. *Earth and Planetary Science Letters*. Vol. 270, pp. 200.

Fletcher, R.C.; Le Craw, R.C.; Spencer, E.G. (1960). Electron spin relaxation in ferromagnetic insulators. *Phys. Rev*. Vol. 117, pp. 955.

Fonseca, F.C.; Goya, G.F.; Jardim, R.F.; Muccillo, R.; Carreño, N.L.V.; Longo, E.; Leite, E.R. (2002). Superparamagnetism and magnetic properties of Ni nanoparticles embedded in SiO_2. *Phys. Rev. B.* Vol. 66, pp.104406.

Frait, Z.; Macfaden, H. (1965). Ferromagnetic resonance in metals frequency dependence. *Phys. Rev. Vol.*139, pp.A1173.

Frait, Z.; Fraitova, D.; Zarubova, N. (1985). Observation of FMR surface spin wave modes in bulk amorphous ferromagnets. *Phys. Stat. Sol.* (b). Vol. 128, pp. 219.

Frait, Z.; Fraitova, D. (1998). Low energy spinwave excitation in highly conducting thin films and surfaces, in *Frontiers in Magnetism of Reduced Dimension Systems.* Nato ASI Series, Bar'yakhtar, V.G.; Wigen, P.E.; Lesnik, N.A. (Eds.) (Kluwer, Dordrecht) pp. 121.

Frait, Z.; Macfaden, H. (1965). Ferromagnetic resonance in metals frequency dependence. *Phys. Rev.* Vol. 139, pp. 1173.

Fraitova, D. (1983(a)). An analytical theory of FMR in bulk metals, I. Dispersion relations. *Phys. Stat. Sol.* Vol. 120, pp. 341.

Fraitova, D. (1983(b)). An analytical theory of FMR in bulk metals, II. Penetration depths. *Phys. Stat. Sol.* Vol. 120, pp. 659.

Fraitova, D. (1984). An analytical theory of FMR in bulk metals, III. Surface impedance. *Phys. Stat. Sol.* Vol. 124, pp. 587.

Fullerton, E.E.; Jiang, J.S.; Bader, S.D. (1999). Hard/soft magnetic heterostructures: model exchange-spring magnets. *J. Magn. Magn. Mater.* Vol. 200, pp. 392.

Fullerton, E.E.; Jiang, J.S.; Grimsditch, M.; Sowers, C.H.; Bader, S.D. (1998). Exchange-spring behavior in epitaxial hard/soft magnetic bilayers. *Phys. Rev. B.* Vol. 58, pp.12193.

Gatteschi, D.; Sessoli, R.; Villain, J. (2006). Molecular nanomagnets. *Oxford University Press.*

García-Miquel, H.; García, J.M.; García-Beneytez, J.M.; Vázquez, M. (2001). Surface magnetic anisotropy in glass-coated amorphous microwires as determined from ferromagnetic resonance measurements. *J. Magn. Magn. Mater.* Vol. 231, pp. 38.

Gerlach, W.; Stern, O. (1922). Dasmagnetische moment dessilber atoms. *Zeitschrift für Physik.* Vol. 9, pp. 353.

Gilbert, T.A. (1955). Armour research foundation rept. *Armour Research Foundation.* Chicago. No.11.

Gnatzig, K.;. Dötsch, H.; Ye, M.; Brockmeyer, A. (1987). Ferrimagnetic resonance in garnet films at large precession angles. *J. Appl. Phys.* Vol. 62, pp. 4839. Golosovsky, M.; Monod, P.; Muduli, P.K.; Budhani, R.C. (2012). Low-field microwave absorption in epitaxial $La_{0.7}Sr_{0.3}MnO_3$ films resulting from the angle-tuned ferromagnetic resonance in the multidomain state. *Arxiv: 1206-3041. Cont-mat. mtrl-sci.* pp. 1.

Goryunov, Yu. V.; Garifyanov, N.N.; Khaliullin, G.G.; Garifullin, I.A.; Tagirov, L.R.; Schreiber, F.; Mühge, Th.; Zabel, H. (1995). Magnetican isotropies of sputtered Fe films on MgO substrates. *Phys. Rev. B.* Vol. 52, pp. 13450.

Grimsditch, M.; Camley, R.; Fullerton, E.E.; Jiang, S.; Bader, S.D.; Sowers, H. (1999). Exchange-spring systems: Coupling of hard and soft ferromagnets as measured by magnetization and Brillouin light scattering (invited). *J. Appl. Phys.* Vol. 85, pp. 5901.

Grünberg, P.; Schwarz, B.; Vach, W.; Zinn, W.; Dabkowski, D. (1979). Light scattering from spin waves in bubble films. *J. Magn. Magn. Mater.* Vol. 13, pp. 181.

Grünberg, P. (1980). Brillouin scattering from spin waves in thin ferromagnetic films. *J. Magn. Magn. Mater.* Vol. 15–18, pp. 766.

Grünberg, P.; Mayr, C.M.; Vach, W. (1982). Determination of magnetic parameters by means of brillouin scattering. Examples: Fe, Ni, $Ni_{0.8}Fe_{0.2}$. *J. Magn. Magn. Mater.* Vol. 28, pp. 319.

Grünberg, P.; Barnas, J.; Saurenbach, F.; Fuβ, J.A.; Wolf, A.; Vohl, M. (1991). Layered magnetic structures: antiferromagnetic type interlayer coupling and magnetoresistance due to antiparallel alignment. *J. Magn. Magn. Mater.* Vol. 93, pp. 58.

Grünberg, P.; Demokritov, S.; Fuss, A.; Schreiber, R.; Wolf, J.A.; Purcell, S.T. (1992). Interlayer exchange, magnetotransport and magnetic domains in Fe/Cr layered structures. *J. Magn. Magn. Mater.* Vol. 104-107, pp. 1734.

Grünberg, P. (2000). Layered magnetic structures in research and application. *Acta mater.* Vol. 48, pp. 239. Grünberg, P. (2001). Layered magnetic structures: history, facts and figures. *J. Magn. Magn. Mater.* Vol. 226-230, pp. 1688.

Guimarães, A.P. (2009). Principles of nanomagnetism. *Springer-Verlag Berlin Heidelberg.* ISBN 978-3-642-01481-9. Guimarães, A.P. (1998). Magnetism and magnetic resonance in solids. *A Wiley-Interscience Publication.* Canada.

Gurevich, A.G.; Melkov, G.A. (1996). *Magnetization Oscillations and Waves* (CRC, Boca Raton).

Güner, S.; Yalçın, O.; Kazan, S.; Yıldız, F.; Şahingöz, R. (2006). FMR studies of bilayer $Co_{90}Fe_{10}/Ni_{81}Fe_{19}$, $Ni_{81}Fe_{19}/Co_{90}Fe_{10}$ and monolayer $Ni_{81}Fe_{19}$ thin films. *Phys. Stat. Solid (a).* Vol. 203, pp. 1539.

Hamoudeh, M.; Al Faraj, A.; Canet-Soulas, E.; Bessueille, F.; Leonard, D.; Fessi, H. (2007). Elaboration of PLLA-based superparamagnetic nanoparticles: Characterization, magnetic behaviour study and in vitro elaxivity evaluation. *Int. J. Pharmaceut.* Vol. 338, pp. 248.

Hathaway, K.; Cullen, J. (1981). Magnetoelastic softening of moduli and determination of magnetic anisotropy in RE-TM compounds. *J. Appl. Phys.* Vol. 52, pp. 2282.

Heim, E.; Harling, S.; Pöhlig, K.; Ludwig, F.; Menzel, H.; Schilling, M. (2007). Fluxgate magnetorelaxometry of superparamagnetic nanoparticles for hydrogel characterization. *J. Magn. Magn. Mater.* Vol. 311, pp. 150.

Heinrich, B. (2005(a)). Ferromagnetic resonance in ultrathin film structures. In magnetic ultrathin films. Heinrich, B.; Bland, J.A.C. (Eds.) *Springer, Berlin, Heidelberg.* Vol. II, pp. 195.

Heinrich, B.; Bland, J.A.C. (2005(b)). Ultrathin magnetic structures IV, applications of nanomagnetism. *Springer Berlin Heidelberg* New York . ISBN 3-540-21954-4.

Heisenberg, W. (1928). Theory of ferromagnetism. *ZeitschriftfürPhysik.* Vol. 49, pp. 619.

Held, G.A.; Grinstein, G.; Doyle, H.; Sun, S.; Murray, C.B. (2001). Competing interactions in dispersions of superparamagnetic nanoparticles. *Phys. Rev. B.* Vol. 64, pp. 012408.

Hellwig, O.; Kortrigh, J.B.; Takano, K.; Fullerton, E.E. (2000). Switching behavior of Fe-Pt/Ni-Fe exchange-
spring films studied by resonant soft-x-ray magneto-optical Kerr effect. *Phys. Rev. B.* Vol. 62, pp. 11694.

Herring, C.; Kittel, C. (1950). On the theory of spin waves in ferromagnetic media. *Phys. Rev.* Vol. 81, pp. 869.

Hillebrands, B. (2000). Light scattering in solids VII. Topics. *Appl. Phys.* Vol. 75, M. Cardona, Güntherodt, G. (Eds.). *Springer, Berlin, Heidelberg.*

Hillebrands, B.; Thiaville, A. (2006). Spin dynamics in confined magnetic structures III. *Springer-Verlag Berlin Heidelberg.* ISBN-10 3-540-20108-4.

Hillebrands, B.; Ounadjela, K. (2002). Spin dynamics in confined magnetic structures I. *Springer-Verlag Berlin Heidelberg.*

Hillebrands, B.; Ounadjela, K. (2003). Spin dynamics in confined magnetic structures II. *Springer-Verlag Berlin Heidelberg.* Hinata, S.; Saito, S.; Takahashi, M. (2012). Ferromagnetic resonance analysis of internal effective field of classified grains by switching field for granular perpendicular recording media. *J. Appl. Phys..* Vol. 111, pp. 07B722.

Hsia, L.C.; Wigen, P.E. (1981). Enhancement of uniaxial anisotropy constant by introducing oxygen vacancies in Ca-doped YIG. *J. Appl. Phys.* Vol. 52, pp. 1261.

Huang, Z.C.; Hu, X.F.; Xu, Y.X.; Zhai, Y.; Xu, Y.B.; Wu, J.; Zhai, H. R. (2012). Magnetic properties of ultrathin single crystal Fe_3O_4 film on InAs(100) by ferromagnetic resonance. *J. Appl. Phys.* Vol. 111, pp. 07C108.

Huang, M.D.; Lee, N.N.; Hyun, Y.H.; Dubowik, J.; Lee, Y.P. (2004). Ferromagnetic resonance study of magnetic-shape-memory Ni_2MnGa films. *J. Magn. Magn. Mater.* Vol. 272–276, pp. 2031.

Jalali-Roudsar, A.A.; Denysenkov, V.P.; Khartsev, S.I. (2005). Determination of magnetic anisotropy constants for magnetic garnet epitaxial films using ferromagnetic resonance. *J. Magn. Magn. Mater.* Vol. 288, pp. 15.

Jarrett, H.S.; Waring, R.K. (1958). Ferrimagnetic resonance in $NiMnO_3$. *Phys. Rev.* Vol. 111, pp. 1223.

Jiang, J.S.; Fullerton, E.E.; Sowers, C.H.; Inomata, A.; Bader, S.D.; Shapiro, A.J.; Shull, R.D.; Gornakov, V. S.; Nikitenko, V.I. (1999). Spring magnet films. *IEEE Trans. Magn.* Vol. 35, pp. 3229.

Jiang, J.S.; Pearson, J.E.; Liu, Z.Y.; Kabius, B.; Trasobares, S.; Miller, D.J.; Bader, S.D. (2005). A new approach for improving exchange-spring magnets. *J. Appl. Phys.* Vol. 97, pp. 10K311.

Jiang, J.S.; Bader, S.D.; Kaper, H.; Leaf, G.K.; Shull, R.D.; Shapiro, A.J.; Gornakov, V.S.; Nikitenko, V.I.; Platt, C.L.; Berkowitz, A.E.; David, S.; Fullerton, E.E. (2002). Rotational hysteresis of exchange-spring Magnets. *J. Phys. D: Appl. Phys.* Vol. 35, pp. 2339.

Jung, S.; Watkins, B.; DeLong, L.; Ketterson, J.B.; Chandrasekhar, V. (2002). Ferromagnetic resonance in periodic particle arrays. *Phys. Rev. B.* Vol. 66, pp. 132401.

Kakazei, G.N.; Kravets, A.F.; Lesnik, N.A.; de Azevedo, M.M.P.; Pogorelov, Y. G.; Sousa, J.B. (1999). Ferromagnetic resonance in granular thin films. *J. Appl. Phys.* Vol. 85, pp. 5654.

Kakazei, G.N.; Pogorelov, Yu.G.; Sousa, J.B.; Golub, V.O.; Lesnik, N.A.; Cardoso, S.; Freitas, P.P. (2001). FMR in $CoFe(t)/Al_2O_3$ multilayers: from continuous to discontinuous regime. *J. Magn. Magn. Mater.* Vol. 226-230, pp. 1828.

Kambe, T.; Kajiyoshi, K.; Oshima, K.; Tamura, M.; Kinoshita, M. (2005). Ferromagnetic resonance in β-p-NPNN at radio-frequency region. *Polyhedron*. Vol. 24, pp. 2468.

Kartopu, G.; Yalçın, O.; Kazan, S.; Aktaş, B. (2009). Preparation and FMR analysis of Co nanowires in alumina templates. *J. Magn. Magn. Mater*. Vol. 321, pp. 1142.

Kartopu, G.; Yalçın, O. (2010). Electrodeposited nanowires and their applications. edited by N. Lupu. *INTECH*. available from: http://sciyo.com/articles/show/title/fabrication-and-applications-of-metalnanowire-arrays-electrodeposited-in-ordered-porous-templates.

Kartopu, G.; Yalçın, O.; Choy, K.-L.; Topkaya, R.; Kazan, S.; Aktaş, B. (2011 (a)). Size effects and origin of easy-axis in nickel nanowire arrays. *J. Appl. Phys*. Vol. 109, pp. 033909.

Kartopu, G; Yalçın, O; Demiray, A.S. (2011 (b)). Magnetic and transport properties of chemical solution deposited (100)-textured $La_{0.7}Sr_{0.3}MnO_3$ and $La_{0.7}Ca_{0.3}MnO_3$ nanocrystalline thin films. *Phys. Scr*. Vol. 83, pp. 015701.

Kasuya, T. (1956). A theory of metallic ferro- and antiferromagnetism on Zener's model. *Prog. Theor. Phys*. Vol. 16, pp. 45.

Kind, J.; van Raden, U.J.; García-Rubio, I.; Gehring, A.U. (2012). Rock magnetic techniques complemented by ferromagnetic resonance spectroscopy to analyse a sediment record. *Geophys. J. Int*. Vol. 191, pp. 51.

Kip, A.F.; Arnold, R.D. (1949). Ferromagnetic resonance at microwave frequencies in iron single crystal. *Phys. Rev*. Vol. 75, pp. 1556.

Kittel, C. (1947). On the theory of ferromagnetic resonance absorption. *Phys. Rev*. Vol. 73, pp. 155.

Kittel, C. (1946).Theory of the structure of ferromagnetic domains in films and small Particles. *Phys. Rev*. Vol. 70, pp. 965.

Kittel, C. (1949). On the gyromagnetic ratio and spectroscopic splitting factor of ferromagnetic substances. *Phys. Rev*. Vol. 76, pp. 743.

Kittel, C.; Abrahams, E. (1953). Relaxation processes in ferromagnetism. *Rev. Mod. Phys*. Vol. 25, pp. 233.

Kip, A.F.; Arnold, R.D. (1949). Ferromagnetic Resonance at Microwave Frequencies in Iron Single Crystal. *Phys. Rev*. Vol. 75, pp. 1556.

Kittel, C. (1958). Interaction of spin waves and ultrasonic waves in ferromagnetic crystals. *Phys. Rev*. Vol. 110, pp. 836.

Kharmouche, A.; Ben Youssef, J.; Layadi, A.; Chérif, S.M. (2007). Ferromagnetic resonance in evaporated Co/Si(100) and Co/glass thin films. *J. App. Phys*. Vol. 101, pp. 13910.

Khaibullin, R.I.; Tagirov, L.R.; Rameev, B.Z.; Ibragimov, S.Z.; Yıldız, F.; Aktaş, B. (2004). High curie-temperature ferromagnetism in cobalt-implanted single-crystalline rutile. *J. Phys.: Condens. Matter*. Vol. 16, pp. 1.

Klein, P.; Varga, R.; Infante, G.; Vázquez, M. (2012). Ferromagnetic resonance study of FeCoMoB microwires during devitrification process. *J. Appl. Phys*. Vol. 111, pp. 053920.

Kliava, J.; Berger, R. (1999). Size and shape distribution of magnetic nanoparticles in disordered systems: computer simulations of superparamagnetic resonance spectra. *J. Magn. Magn. Mater*. Vol. 205, pp. 328.

Kneller, E.F.; Hawing, R. (1991). The exchange-spring magnet: a new material principle for permanent magnets. *IEEE Trans. Magn*. Vol. 27, pp. 3588.

Kobayashi, T.; Ishida, N.; Sekiguchi, K.; Nozaki, Y. (2012). Ferromagnetic resonance properties of granular Co-Cr-Pt films measured by micro-fabricated coplanar waveguides. *J. Appl. Phys.* Vol. 111, pp. 07B919.

Kodama, R.H. (1999). Magnetic Nanoparticles. *J. Magn. Magn. Mater.* Vol. 200, pp. 359.

Kodama, R.H.; Berkowitz, A.E. (1999). Atomic-scale magnetic modeling of oxide nanoparticles. *Phys. Rev. B.* Vol. 59, pp. 6321.

Kodama, R.H.; Berkowitz, A.E.; McNiff Jr. E.J.; Foner S. (1996). Surface spin disorder in $NiFe_2O_4$ nanoparticles. *Phys. Rev. Lett.* Vol. 77, pp. 394.

Kooi, C.F.; Wigen, P.E.; Shanabarger, M.R.; Kerrigan J.V. (1964). Spin-wave resonance in magnetic films on the basis of surface –spin –pinning model and the volume inhomogeneity model. *J. Appl. Phys.* Vol. 35, pp. 791.

Kopp, R.E.; Weiss, B.P.; Maloof, A.C.; Vali, H.; Nash, C.Z.; Kirschvink, J.L. (2006). Chains, clumps, and strings: magneto fossil taphonomy with ferromagnetic resonance spectroscopy. *Earth Planet Sc. Lett.* Vol. 247, pp. 10.

Korolev, K.A.; McCloy, J.S.; Afsar, M.N. (2012). Ferromagnetic resonance of micro- and nano-sized hexagonal ferrite powders at millimeter waves. *J. Appl. Phys.* Vol. 111, pp. 07E113.

Kohmoto, O. (2007). Ferromagnetic resonance equation of hexagonal ferrite in c-plane. *J. Magn. Magn. Mater.* Vol. 310, pp. 2561.

Knorr, T.G.; Hoffman, R.W. (1959). Dependence of geometric magnetic anisotropy in thin iron films. *Phys. Rev.* Vol. 113, pp. 1039.

Kraus, L.; Frait, Z.; Ababei, G.; Chayka, O.; Chiriac, H. (2012). Ferromagnetic resonance in submicron amorphous wires. *J. Appl. Phys.* Vol. 111, pp. 053924.

Krebs, J.J.; Rachford, F.J.; Lubitz, P.; Prinz, G.A. (1982). Ferromagnetic resonance studies of very thin epitaxial single crystals of iron. *J. Appl. Phys.* Vol. 53, pp. 8058.

Krebs, J.J.; Lubitz, P.; Chaiken, A.; Prinz, G.A. (1989). Magnetic resonance determination of the antiferromagnetic coupling of Fe layers through Cr. *Phys. Rev. Lett.* Vol. 63, pp. 1645.

Krivoruchko, V.N.; Marchenko, A.I. (2012). Spatial confinement of ferromagnetic resonances in honeycomb antidot lattices. *J. Magn. Magn. Mater.* Vol. 324, pp. 3087.

Krone, P.; Albrecht, M.; Schrefl, T. (2011). Micromagnetic simulation of ferromagnetic resonance of perpendicular granular media: Influence of the intergranular exchange on the Landau–Lifshitz–Gilbert damping constant. *J. Magn. Magn. Mater.* Vol. 323, pp. 432.

Kuanr, B.K.; Camley, R.E.; Celinski, Z. (2004). Relaxation in epitaxial Fe films measured by ferromagnetic resonance. *J. Appl. Phys.* Vol. 93, pp. 6610.

Kuanr, B.K.; Camley, R.E.; Celinski, Z. (2005). Extrinsic contribution to Gilbert damping in sputtered NiFe films by ferromagnetic resonance. *J. Magn. Magn. Mater.* Vol. 286, pp. 276.

Kudryavtsev, Y.V.; Oksenenko, V.A.; Kulagin, V.A.; Dubowik, J.; Lee, Y.P. (2007). Ferromagnetic resonance in Co2MnGa films with various structural ordering. *J. Magn. Magn. Mater.* Vol. 310, pp. 2271.

Lacheisserie, E.T.; Gignoux, D.; Schlenker, M. (2005). Magnetism, materials and aplications. *Springer science + Business Media, Inc.* Boston.

Landau, E.; Lifshitz, E. (1935). On the theory of the dispersion of magnetic permeability in ferromagnetic bodies. *Physik Z. Sowjetunnion.* Vol. 8, pp. 153. Layadi, A.; Lee, J.M.; Artman, J.O. (1988). Spin-wave FMR in annealed NiFe/FeMn thin films. *J. Appl. Phys.* Vol. 63, pp. 3808.

Layadi, A. (2002). Exchange anisotropy: A ferromagnetic resonance study. *Phys. Rev. B.* Vol. 66, pp. 184423.

Layadi, A. (2004). Theoretical study of resonance modes of coupled thin films in the rigid layer model. *Phys. Rev. B.* Vol. 69, pp. 144431.

Layadi, A.; Artman, J.O. (1990(a)). Ferromagnetic resonance in a coupled two-layer System. *J. Magn. Magn. Mater.* Vol. 92, pp. 143.

Layadi A.; Artman, J.O. (1990(b)). Study of antiferromagnetic coupling by ferromagnetic resonance (FMR).*J. Magn. Magn. Mater.* Vol. 92, pp. 143.

Lee, J.; Hong, Y.K.; Lee, W.; Abo, G.S.; Park, J.; Syslo, R.; Seong, W.M.; Park, S.H.; Ahn, W.K. (2012). High ferromagnetic resonance and thermal stability spinel $Ni_{0.7}Mn_{0.3-x} Co_xFe_2O_4$ ferrite for ultra high frequency devices. *J. Appl. Phys.* Vol. 111, pp. 07A516.

Limmer, W.; Glunk, M.; Daeubler, J.; Hummel, T.; Schoch, W.; Bihler, C.; Huebl, H.; Brandt, M.S.; Goennenwein, S.T.B.; Sauer, R. (2006). Magnetic anisotropy in (Ga, Mn)As on GaAs(1 1 3)As studied by magnetotransport and ferromagnetic resonance. *Microelectron. J.* Vol. 37, pp. 1490.

Lindner, J.; Baberschke, K. (2003). In situ ferromagnetic resonance: an ultimate tool to investigate the coupling in ultrathin magnetic films. *J. Phys.: Condens. Matter.* Vol. 15, pp. R193.

Lindner, J.; Tolinski, T.; Lenz, K.; Kosubek, E.; Wende, H.; Baberschke, K.; Ney, A.; Hesjedal, T.; Pampuch, C.; Koch, R.; Däweritz, L.; Ploog, K.H. (2004). Magnetic anisotropy of MnAs-films on GaAs(0 0 1) studied with ferromagnetic resonance. *J. Magn. Magn. Mater.* Vol. 277, pp. 159.

Li, N.; Schäfer, S.; Datta, R.; Mewes, T.; Klein, T. M.; Gupta, A. (2012). Microstructural and ferromagnetic resonance properties of epitaxial nickel ferrite films grown by chemical vapor deposition. *Appl. Phys. Lett.* Vol. 101, pp. 132409.

Liua, B.; Yang, Y.; Tang, D.; Zhang, B.; Lu, M.; Lu, H. (2012). The contributions of intrinsic damping and two magnon scattering on the ferromagnetic resonance linewidth in $[Fe_{65}Co_{35}/SiO_2]_n$ multilayer films. J. Alloy. compd. Vol. 524, pp. 69.

Macdonald, J.R. (1956). Spin exchange effects in ferromagnetic resonance. *Phys. Rev.* Vol. 103, pp. 280.

Maciá, F.; Warnicke, P.; Bedau, D.; Im, M.Y.; Fischer, P.; Arena, D.A.; Kent, A.D. (2012). Perpendicular magnetic anisotropy in ultrathin Co_9Ni multilayer films studied with ferromagnetic resonance and magnetic x-ray microspectroscopy. *J. Magn. Magn. Mater.* Vol. 324, pp. 3629.

Maklakov, S.S.; Maklakov, S.A.; Ryzhikov, I.A.; Rozanov, K.N.; Osipov, A.V. (2012). Thin Co films with tunable ferromagnetic resonance frequency. *J. Magn. Magn. Mater.* Vol. 324, pp. 2108.

Maksymowich, L.J.; Sendorek, D. (1983). Surface modes in magnetic thin amorphous films of GdCoMo alloys. *J. Magn. Magn. Mater.* Vol. 37, pp. 177.

Maksymowicz, L.J.; Sendorek, D.; Żuberek, R. (1985). Surface anisotropy energy of thin amorphous magnetic films of (Gd1-xCox)1-yMoy alloys; experimental results. *J. Magn. Magn. Mater.* Vol. 46, pp. 295. Maksymowicz, L.J.; Jankowski, H. (1992). FMR experiment in multilayer structure of FeBSi/Pd. *J. Magn. Magn. Mater.* Vol. 109, pp. 341.

Marin, C.N. (2006). Thermal and particle size distribution effects on the ferromagnetic resonance in magnetic fluids. *J. Magn. Magn. Mater.* Vol. 300, pp. 397.

Mastrogiacomo, G.; Fischer H.; Garcia-Rubio, I.; Gehring, A.U. (2010). Ferromagnetic resonance spectroscopic response of magnetite chains in a biological matrix. *J. Magn. Magn. Mater.* Vol. 322, pp. 661.

Mazur, P.; Mills, D.L. (1982). Inelastic scattering of neutrons by surface spin waves on ferromagnets. *Phys. Rev. B.* Vol. 26, pp. 5175. McMichael, R.D.; Wigen, P.E. (1990). High power FMR without a degenerate spin wave manifold. *Phys. Rev. Lett.* Vol. 64, pp. 64.

Meiklejohn, W.H.; Bean, C.P. (1956). New magnetic anisotropy. *Phys. Rev.* Vol. 102, pp. 1413.

Meiklejohn, W.H.; Bean, C.P. (1957). New magnetic anisotropy. *Phys. Rev.* Vol. 105, pp. 904.

Mercereau, J.E.; Feynman, R.P. (1956). Physical conditions for ferromagnetic resonance. *Phys. Rev.* Vol. 104, pp. 63.

Mibu, K.; Nagahama, T.; Ono, T.; Shinjo, T. (1997). Magnetoresistance of quasi-Bloch-wall induced in NiFe/CoSm exchange. *J. Magn. Magn. Mater.* Vol. 177-181, pp. 1267.

Mibu, K.; Nagahama, T.; Shinjo, T.; Ono, T. (1998). Magnetoresistance of Bloch-wall-type magnetic structures induced in NiFe/CoSm exchange-spring bilayers. *Phys. Rev. B.* Vol. 58, pp. 6442.

Mills, D.L.; Bland, J.A.C. (2006). Nanomagnetism ultrathin films, multilayers and nanostructures. *Elsevier B.V.* Mills, D.L. (2003). Ferromagnetic resonance relaxation in ultrathin metal films: The role of the conduction electrons. *Phys. Rev. B.* Vol. 68, pp. 014419.

Min, J.H.; Cho, J.U.; Kim, Y.K.; Wu, J.H.; Ko, Y.D.; Chung, J.S. (2006). Substrate effects on microstructure and magnetic properties of electrodeposited Co nanowire arrays. *J. Appl. Phys.* Vol. 99, pp. 08Q510.

Miyake, K.; Noh, S.M.; Kaneko, T.; Imamura, H.; Sahashi, M. (2012). Study on high-frequency 3–D magnetization precession modes of circular magnetic nano-dots using coplanar wave guide vector network analyzer ferromagnetic resonance. *IEEE T. Magn.* Vol. 48, pp. 1782.

Mizukami, S.; Nagashima, S.; Yakata, S.; Ando, Y.; Miyazaki, T. (2007). Enhancement of DC voltage generated in ferromagnetic resonance for magnetic thin film. *J. Magn. Magn. Mater.* Vol. 310, pp. 2248.

Morrish, A. H. (1965).The physical principles of magnetism *Wiley*, New York.

Nagamine, L.C.C.M.; Geshev, J.; Menegotto, T.; Fernandes, A.A.R.; Biondo, A.; Saitovitch, E.B. (2005). Ferromagnetic resonance and magnetization studies in exchange-coupled NiFe/Cu/NiFe structures. *J. Magn. Magn. Mater.* Vol. 288, pp. 205.

Nakai, T.; Yamaguchi, M.; Kikuchi, H.; Iizuka, H.; Arai, K.I. (2002). Remarkable improvement of sensitivity for high-frequency carrier-type magnetic field sensor with ferromagnetic resonance. *J. Magn. Magn. Mater.* Vol. 242–245, pp. 1142.

Neudecker, I.; Woltersdorf, G.; Heinrich, B.; Okuno, T.; Gubbiotti, G.; Back, C.H. (2006). Comparison of frequency, field, and time domain ferromagnetic resonance methods. *J. Magn. Magn. Mater.* Vol. 307, pp. 148.

Olive, E.; Lansac, Y.; Wegrowe, J.E. (2012). Beyond ferromagnetic resonance: The inertial regime of the magnetization. *Appl. Phys. Lett.* Vol. 100, pp. 192407.

Osthöver, C.; Grünberg, P.; Arons, R.R. (1998). Magnetic properties of doped $La_{0.67}Ba_{0.33}Mn_{1-y}A_yO_3$, A - Fe, Cr. *J. Magn. Magn. Mater.* Vol. 177-181, pp. 854.

Owens, F.J. (2009). Ferromagnetic resonance observation of a phase transition in magnetic field-aligned Fe_2O_3 nanoparticles. *J. Magn. Magn. Mater.* Vol. 321, pp. 2386.

Öner, Y.; Özdemir, M.; Aktaş, B.; Topacli, C.; Haris, E.A.; Senoussi, S. (1997). The role of Pt impurities on both bulk and surface anisotropies in amorphous NiMn films. *J. Magn. Magn. Mater.* Vol. 170, pp. 129.

Özdemir, M.; Aktaş, B.; Öner, Y.; Sato, T.; Ando, T. (1996). A spin- wave resonance study on reentrant $Mn_{77}Mn_{23}$ thin films. *J. Magn. Magn. Mater.* Vol. 164, pp. 53.

Özdemir, M.; Aktaş, B.; Öner, Y.; Sato, T.; Ando, T. (1997). Anomalous anisotropy of re-entrant $Ni_{77}Mn_{23}$ film. *J. Phys.: Condens. Matter.* Vol. 9, pp. 6433.

Özdemir, M.; Öner, Y.; Aktaş, B. (1998). Evidence of superparamagnetic behaviour in an amorphous $Ni_{62}Mn_{38}$ film by ESR measurements. *Phys. B.* Vol. 252, pp. 138.

Patel, R.; Owens, F.J. (2012). Ferromagnetic resonance and magnetic force microscopy evidence for above room temperature ferromagnetism in Mn doped Si made by a solid state sintering process. *Solid State Commun.* Vol. 152, pp. 603.

Patton, C.E.; Hurben, M.J. (1995). Theory of magnetostatic waves for in-plane magnetized isotropic films. *J. Magn. Magn. Mater.* Vol. 139, pp. 263. Patton, C.E.; Hurben, M.J. (1996). Theory of magnetostatic waves for in-plane magnetized anisotropic films. *J. Magn. Magn. Mater.* Vol. 163, pp. 39.

Park, D.G.; Kim, C.G.; Kim, W.W.; Hong, J.H. (2007). Study of GMI-valve characteristics in the Co-based amorphous ribbon by ferromagnetic resonance. *J. Magn. Magn. Mater.* Vol. 310, pp. 2295.

Parkin, S.S.P.; More, N.; Roche, K.P. (1990). Oscillations in exchange coupling and magneto resistance in metallic superlattice structures: Co/Ru, Co/Cr, and Fe/Cr. *Phys. Rev. Lett.* Vol. 64, pp. 2304.

Parkin, S.S.P.; Bhadra, R.; Roche, K.P. (1991(a)). Oscillatory magnetic exchange coupling through thin copper layers. *Phys. Rev. Lett.* Vol. 66, pp. 2152.

Parkin, S.S.P. (1991(b)). Systematic variation of the strength and oscillation period of indirect magnetic exchange coupling through the 3*d*, 4*d*, and 5*d* transition metals. *Phys. Rev. Lett.* Vol. 67, pp. 3598.

Parkin, S.S.P.; Farrow, R.F.C.; Marks, R.F.; Cebollada, A.; Harp, G.R.; Savoy, R.J. (1994). Oscillations of interlayer exchange coupling and giant magnetoresistance in (111) oriented permalloy/Au multilayers. *Phys. Rev. Lett.* Vol. 72, pp. 3718.

Parvatheeswara, R.B.; Caltun, O.; Dumitru, I.; Spinu, L. (2006). Ferromagnetic resonance parameters of ball-milled Ni–Zn ferrite nanoparticles. *J. Magn. Magn. Mater.* Vol. 304, pp. 752.

Paul, A.; Bürgler, D.E.; Grünberg, P. (2005). Enhanced exchange bias in ferromagnet/ antiferromagnet multilayers. *J. Magn. Magn. Mater.* Vol. 286, pp. 216.

Paz, E.; Cebollada, F.; Palomares, F.J.; González, J.M.; Martins, J.S.; Santos, N.M., Sobolev, N.A. (2012). Ferromagnetic resonance and magnetooptic study of submicron epitaxial Fe(001) stripes. *J. Appl. Phys.* Vol. 111, pp. 123917.

Pires, M.J.M.; Mansanares, A.M., da Silva, E.C.; Schmidt, J.E.; Meckenstock, R.; Pelzl, J. (2006). Ferromagnetic resonance studies in granular CoCu codeposited films. *J. Magn. Magn. Mater.* Vol. 300, pp. 382.

Platow, W.; Anisimov, A.N.; Dunifer, G.L.; Farle, M.; Baberschke, K. (1998). Correlations between ferromagnetic-resonance linewidths and sample quality in the study of metallic ultrathin films. *Phys. Rev. B.* Vol. 58, pp. 5611.

Pollmann, J.; Srajer, G.; Hakel, D.; Lang, J.C.; Maser, J.; Jiang, J.S.; Bader, S.D. (2001). Magnetic imaging of a buried SmCo layer in a spring magnet. *J. Appl. Phys.* Vol. 89, pp. 7165.

Poole, C.P.; Owens J.F. (2003). Introduction to nanotechnology. *John Wiley & Sons, Inc.* ISBN: 0-471-07935-9.

Prieto, A.G.; Fdez-Gubieda, M.L.; Lezama, L.; Orue, I. (2012). Study of surface effects on CoCu nanogranular alloys by ferromagnetic resonance. *J. Appl. Phys.* Vol. 111, pp. 07C105.

Purcell, E.M.; Torrey, H.C.; Pound, R.V. (1946). Resonance absorption by nuclear magnetic moments in a solid. Phys. Rev. Vol. 69, pp. 37.

Puszkarski, H. (1992). Spectrum of interface coupling-affected spin-wave modes in ferromagnetic bilayer films. *Phys. stat. sol.* Vol. 171, pp. 205.

Rabi, I.I.; Zacharias, J.R.; Millman, S.; Kusch, P. (1938). A new method of measuring nuclear magnetic moment, *Phys. Rev.* Vol. 53, pp. 318.

Rachford, F.J.; Vittoria, C. (1981). Ferromagnetic anti-resonance in non-saturated magnetic metals. *J. Appl. Phys.* Vol. 52, pp. 2253.

Rado, G.T. (1958). Effect of electronic mean free path on spin-wave resonance in ferromagnetic metals. *J. Appl. Phys.* Vol. 29, pp. 330. Raikher, Yu.L.; Stepanov, V.I. (2003). Nonlinear dynamic response of superparamagnetic nanoparticles. *Microelectron. Eng.* Vol. 69, pp. 317.

Raita, O.; Popa, A.; Stan, M.; Suciu, R.C.; Biris, A.; Giurgiu, L.M. (2012). Effect of Fe concentration in ZnO powders on ferromagnetic resonance spectra. *Appl. Magn. Reson.* Vol. 42, pp. 499.

Rahman, F. (2008). Nanostructures in electronics and photonics. *Pan Stanford Publishing Pte. Ltd.* Singapore, 596224.

Rameev, B.Z.; Aktaş, B.; Khaibullin, R.I.; Zhikharev, V.A.; Osin, Yu.N.; Khaibullin, I.B. (2000). Magnetic properties of iron-and cobalt-implanted silicone polymers. *Vacuum.* Vol. 58, pp. 551.

Rameev, B.Z.; Yıldız, F.; Aktaş, B.; Okay, C.; Khaibullin, R.I.; Zheglov, E.P.; Pivin, J.C.; Tagirov, L.R. (2003(a)). I on synthesis and FMR studies of iron and cobalt nanoparticles in polyimides. *Microelectron. Eng.* Vol. 69, pp. 330.

Rameev, B.Z.; Yilgin, R.; Aktaş, B.; Gupta, A.; Tagirov, L.R. (2003(b)). FMR studies of CrO epitaxial thin films. *Microelectron. Eng.* Vol. 69, pp. 336.

Rameev, B.Z.; Gupta, A.; Anguelouch, A.; Xiao, G.; Yıldız, F.; Tagirov, L.R.; Aktaş, B. (2004 (b)). Probing magnetic anisotropies in half-metallic CrO_2 epitaxial films by FMR. *J. Magn. Magn. Mater.* Vol. 272–276, pp. 1167.

Rameev, B.; Okay, C.; Yıldız, F.; Khaibullin, R.I.; Popok, V.N.; Aktas, B. (2004 (a)). Ferromagnetic resonance investigations of cobalt-implanted polyimides. *J. Magn. Magn. Mater.* Vol. 278, pp. 164.

Ramesh, M.; Wigen, P.E. (1988 (a)). Ferromagnetic resonance of parallel stripe domains–domain wall system. *J. Magn. Magn. Mater.* Vol. 74, pp. 123. Ramesh, M.; Ren, E.W.; Artman, J.O.; Kryder, M.H. (1988 (b)). Domain mode ferromagnetic resonance studies in bismuth-substituted magnetic garnet films. *J. Appl. Phys.* Vol. 64, pp. 5483.

Ramprasad, R.; Zurcher P.; Petras, M.; Miller, M.; Renaud, P. (2004). Magnetic properties of metallic ferromagnetic nanoparticle composites. *J. Appl. Phys.* Vol. 96, pp. 519.

Ramsey, N.F. (1985). Molecular beams. *Oxford University Press.* New York.

Raposo, V.; Zazo, M.; Iñiguez, J. (2011). Comparison of ferromagnetic resonance between amorphous wires and microwires. *J. Magn. Magn. Mater.* Vol. 323, pp. 1170.

Reich, K.H. (1955). Ferromagnetic resonance absorption in a nickel single crystal at low temperature. *Phys. Rev.* Vol. 101, pp. 1647.

Rezende, S.M.; Moura, J.A.S.; de Aguiar, F.M. (1993). Ferromagnetic resonance in Ag coupled Ni films. *J. Appl. Phys.* Vol. 73, pp. 6341.

Riedling, S.; Knorr, N.; Mathieu, C.; Jorzick, J.; Demokritov, S.O.; Hillebrands, B.; Schreiber, R.; Grünberg, P. (1999). Magnetic ordering and anisotropies of atomically layered Fe/Au(001) multilayers. *J. Magn. Magn. Mater.* Vol.198-199, pp. 348.

Richard, C.; Houzet, M.; Meyer, J.S. (2012). Andreev current induced by ferromagnetic resonance. *Appl. Phys. Lett.* Vol. 109, pp. 057002.

Rivkin, K.; Xu, W., De Long, L.E.; Metlushko, V.V.; Ilic, B.; Ketterson, J.B. (2007). Analysis of ferromagnetic resonance response of square arrays of permalloy nanodots. *J. Magn. Magn. Mater.* Vol. 309, pp. 317.

Rodbell, D.S. (1964). Ferromagnetic resonance absorption linewidth of nickel metal. Evidence for Landau- Lifshitz damping. *Phys. Rev. Lett.* Vol. 13, pp. 471.

Rook, K.; Artman, J.O. (1991). Spin wave resonance in FeAlN films. *IEEE Trans. Magn.* Vol. 27, pp. 5450. Roy, W.V.; Boeck, J.D.; Borghs, G. (1992). Optimization of the magnetic field of perpendicular ferromagnetic thin films for device applications. *Appl. Phys. Lett.* Vol. 61, pp. 3056.

Rousseau, O.; Viret, M. (2012). Interaction between ferromagnetic resonance and spin currents in nanostructures. *Phys. Rev. B.* Vol. 85, pp. 144413.

Römer, F.M.; Möller, M.; Wagner, K.; Gathmann, L.; Narkowicz, R.; Zähres, H.; Salles, B.R.; Torelli, P.; Meckenstock, R.; Lindner, J.; Farle, M. (2012). In situ multifrequency ferromagnetic resonance and x-ray magnetic circular dichroism investigations on Fe/GaAs(110): Enhanced g-factor. *J. Appl. Phys.* Vol. 100, pp. 092402.

Ruderman, M.A.; Kittel, C. (1954). Indirect exchange coupling of nuclear magnetic moments by conduction electrons. *Phys. Rev.* Vol. 96, pp. 99. Rusek, P. (2004). Spin dynamics of ferromagnetic spin glass. *J. Magn. Magn. Mater.* Vol. 272–276, pp. 1332.

Sarmiento, G.; Fdez-Gubieda, M.L.; Siruguri, V.; Lezama, L.; Orue, I. (2007). Ferromagnetic resonance study of $Fe_{50}Ag_{50}$ granular film. *J. Magn. Magn. Mater.* Vol. 316, pp. 59.

Sasaki, M.; Jönsson, P.E.; Takayama, H.; Mamiya, H. (2005). Aging and memory effects in superparamagnets and superspin glasses. *Rev. B.* Vol. 71, pp. 104405.

Schäfer, S.; Pachauri, N.; Mewes, C.K.A.; Mewes, T.; Kaiser, C.; Leng, Q.; Pakala, M. (2012). Frequency-selective control of ferromagnetic resonance linewidth in magnetic multilayers. *J. Appl. Phys.* Vol. 100, pp. 032402.

Schmool, D.S.; Barandiarán, J.M. (1998). Ferromagnetic resonance and spin wave resonance in multiphase materials: theoretical considerations. *J. Phys.: Condens. Matter.* Vol. 10, pp. 10679.

Schmool, D.S.; Schmalzl, M. (2007). Ferromagnetic resonance in magnetic nanoparticle assemblies. *J. Non-Cryst. Solids.* Vol. 353, pp. 738.

Scholz, W.; Suess, D.; Schrefl, T.; Fidler, J. (2000). Micromagnetic simulation of structure-property relations in hard and soft magnets. *Comp. Mater. Sci.* Vol. 18, pp. 1.

Schreiber, F.; Frait, Z. (1996). Spinwave resonance in high conductivity films: The Fe–Co alloy system. *Phys. Rev. B.* Vol. 54, pp. 6473. Schrefl, T.; Kronmüller, H.; Fidler, J. (1993(a)). Exchange hardening in nano-structured permanent magnets. *J. Magn. Magn. Mater.* Vol. 127, pp. 273.

Schrefl, T.; Schmidts, H.F.; Fidler, J.; Kronmüller, H. (1993(b)). The role of exchange and dipolar coupling at grain boundaries in hard magnetic materials. *J. Magn. Magn. Mater.* Vol. 124, pp. 251. Schrefl, T.; Fidler, J. (1998). Modelling of exchange-spring permanent magnets. *J. Magn. Magn. Mater.* Vol. 177-181, pp. 970.

Schrefl, T.; Forster, H.; Fidler, J.; Dittrich, R.; Suess, D.; Scholz, W. (2002). Magnetic hardening of exchange spring multilayers. Proof XVII Rare Earth Magnets Workshop, University of Delaware, ed: Hadjipanayis, G.; Bonder M.J. pp. 1006.

Schultz, S.; Gullikson, E.M. (1983). Measurement of static magnetization using electron spin resonance. *Rev. Sci. Insrum.* Vol. 54, pp. 1383.

Seemann, K.; Leiste, H.; Klever, C. (2009). On the relation between the effective ferromagnetic resonance linewidth Δf_{eff} and damping parameter α_{eff} in ferromagnetic Fe–Co–Hf–N nanocomposite films. *J. Magn. Magn. Mater.* Vol. 321, pp. 3149.

Seemann, K.; Leiste, H.; Klever, Ch. (2010). Determination of intrinsic FMR line broadening in ferromagnetic $(Fe_{44}Co_{56})_{77}Hf_{12}N_{11}$ nanocomposite films. *J. Magn. Magn. Mater.* Vol. 322, pp. 2979.

Seib, J.; Steiauf, D.; Fähnle, M. (2009).Linewidth of ferromagnetic resonance for systems with anisotropic damping. *Phys. Rev. B.* Vol. 79, pp. 092418.

Sellmyer, D.; Skomski, R. (2006). Advanced magnetic nanostructures. *Springer Science+Business Media, Inc.*

Sihues, M.D.; Durante-Rincón, C.A.; Fermin, J.R. (2007). A ferromagnetic resonance study of NiFe alloy thin films. *J. Magn. Magn. Mater.* Vol. 316, pp. 462.

Singh, A.; Chowdhury, P.; Padma, N.; Aswal, D.K.; Kadam, R.M.; Babu, Y.; Kumar, M.L.J.; Viswanadham, C.S.; Goswami, G.L.; Gupta, S.K.; Yakhmi, J.V. (2006). Magneto-transport and ferromagnetic resonance studies of polycrystalline $La_{0.6}Pb_{0.4}NO_3$ thin films. *Solid State Commun.* Vol. 137, pp. 456.

Shames, A.I.; Rozenberg, E.; Sominski, E.; Gedanken, A. (2012). Nanometer size effects on magnetic order in La1-x CaxMnO3 (x = 50.5 and 0.6) manganites, probed by ferromagnetic resonance. *J. Appl. Phys.* Vol. 111, pp. 07D701.

Sklyuyev, A.; Ciureanu, M.; Akyel, C.; Ciureanu, P.; Yelon, A. (2009). Microwave studies of magnetic anisotropy of Co nanowire arrays. *J. App. Phys.* Vol. 105, pp. 023914.

Skomski, R.; Coey, J.M.D. (1993). Giant energy product in nanostructured two-phase magnets. *Phys. Rev. B.* Vol. 48, pp. 15812.

Skomski, R. (2008) .Simple models of magnetism. *Oxford University Press.* ISBN 978-0-19-857075-2.

Slichter, C.P. (1963). Principle of Magnetic Resonance. *Harper & Row,* New York. Smit, J.; Beljers, H. G. (1955). Ferromagnetic resonance absorption in $BaFe_{12}O_{19}$, a high anisotropy crystal. *Philips Res. Rep.* Vol. 10, pp. 113.

Spaldin, N.A. (2010). Magnetic materials: Fundamentals and applications. *Cambridge University Press.* ISBN 13 978 0 521 88669 7.

Sparks, M. (1964). Ferromagnetic Relaxation Theory. *McGraw-Hill,* New York.

Sparks, M. (1969). Theory of surface spin pinning in ferromagnetic resonance. *Phys. Rev. Lett.* Vol. 22, pp. 1111.

Sparks, M. (1970(a)). Ferromagnetic resonance in thin films. III. Theory of mode Intensities. *Phys. Rev. B.* Vol. 1, pp. 3869.

Sparks, M. (1970(b)). Ferromagnetic resonance in thin films I and II. *Phys. Rev. B.* Vol. 1, pp. 3831. Sparks, M. (1970(c)). Ferromagnetic resonance in thin films. I. Theory of normal-mode frequencies. *Phys. Rev. B.* Vol. 1, pp. 3831.

Sparks, M. (1970(d)). Ferromagnetic resonance in thin films. I. Theory of normal-mode intensities. *Phys. Rev. B.* Vol. 1, pp. 3869.

Speriosu, V.S.; Parkin, S.S.P. (1987). Standing spin waves in FeMn/NiFe/FeMn exchange bias structures. *IEEE Trans. magn.* Vol. 23, pp. 2999.

Spinu, L.; Dumitru, I.; Stancu, A.; Cimpoesu, D. (2006). Transverse susceptibility as the low-frequency limit of ferromagnetic resonance. *J. Magn. Magn. Mater.* Vol. 296, pp. 1.

Spoddig, D.; Meckenstock, R.; Bucher, J.P.; Pelzl, J. (2005). Studies of ferromagnetic resonance line width during electrochemical deposition of Co films on Au(1 1 1). *J. Magn. Magn. Mater.* Vol. 286, pp. 286.

Song, Y.Y.; Kalarickal, S.; Patton, C.E. (2003). Optimized pulsed laser deposited barium ferrite thin films with narrow ferromagnetic resonance linewidths. *J. Appl. Phys.* Vol. 94, pp. 5103.

Song, H.; Mulley, S.; Coussens, N.; Dhagat, P.; Jander, A.; Yokochi, A. (2012). Effect of packing fraction on ferromagnetic resonance in $NiFe_2O_4$ nanocomposites. *J. Appl. Phys.* Vol. 111, pp. 07E348.

Stashkevich, A.A.; Roussigné, Y.; Djemia, P.; Chérif, S.M.; Evans, P.R.; Murphy, A.P.; Hendren, W.R.; Atkinson, R.; Pollard, R.J.; Zayats, A.V.; Chaboussant, G.; Ott, F. (2009).

Spin-wave modes in Ni nanorod arrays studied by Brillouin light scattering. *Phys. Rev. B.* Vol. 80, pp. 144406.

Stoner, E.C.; Wohlfarth, E. P. (1948). A mechanism of magnetic hysteresis in heterogeneous alloys. *Phil. Trans. R. Soc. A.* Vol. 240, pp. 599.

Sun, Y.; Duan, L.; Guo, Z.; DuanMu, Y.; Ma, M.; Xu, L.; Zhang, Y.; Gu, N. (2005). An improved way to prepare superparamagnetic magnetite-silica core-shell nanoparticles for possible biological application. *J. Magn. Magn. Mater.* Vol. 285, pp. 65.

Sun, Y.; Song, Y.Y.; Chang, H.; Kabatek, M.; Jantz, M.; Schneider, W.; Wu, M.; Schultheiss, H.; Hoffmann, A. (2012(a)). Growth and ferromagnetic resonance properties of nanometer-thick yttrium iron garnet films. *Appl. Phys. Lett.* Vol. 101, pp. 152405.

Sun, Y.; Song Y.Y.; Wu, M. (2012(b)). Growth and ferromagnetic resonance of yttrium iron garnet thin films on metals. *Appl. Phys. Lett.* Vol. 101, pp. 082405.

Sun, L.; Wang, Y.; Yang, M.; Huang, Z.; Zhai, Y.; Xu, Y.; Du, J.; Zhai, H. (2012(c)). Ferromagnetic resonance studies of Fe thin films with dilute heavy rare-earth impurities. *J. Appl. Phys.* Vol. 111, pp. 07A328.

Suzuki, Y.; Katayama, T.; Takanashi, K.; Schreiber, R.; Gtinberg, P.; Tanaka, K. (1997) . The magneto-optical effect of Cr(001) wedged ultrathin films grown on Fe(001). *J. Magn. Magn. Mater.* Vol. 165, pp. 134.

Szlaferek, A. (2004). Model exchange-spring nanocomposite. *Status Solidi B,* Vol. 241, pp. 1312.

Tannenwald, P.E.; Seawey, M.H. (1957). Ferromagnetic resonance in thin films of permalloy. *Phys. Rev.* Vol. 105, pp. 377.

Tartaj, P.; González-Carreño, T.; Bomati-Miguel, O.; Serna, C. J. (2004). Magnetic behavior of superparamagnetic Fe nanocrystals confined inside submicron-sized spherical silica particles. *Phys. Rev. B.* Vol. 69, pp. 094401.

Teale, R.W.; Pelegrini, F. (1986). Magnetic surface anisotropy and ferromagnetic resonance in the single crystal GdAl2. *J. Phys. F:Met. Phys.* Vol. 16, pp. 621.

Terry, E.M. (1917). The magnetic properties of iron, nickel and cobalt above the curie point, and Keeson's quantum theory of magnetism. *Phys. Rev.* Vol. 9, pp. 394.

Topkaya, R.; Erkovan, M.; Öztürk, A.; Öztürk, O.; Aktaş, B.; Özdemir, M. (2010). Ferromagnetic resonance studies of exchange coupled ultrathin Py/Cr/Py trilayers. *J. Appl. Phys.* Vol. 108, pp. 023910.

Tsai, C.C.; Choi, J.; Cho, S.; Lee, S.J.; Sarma, B.K.; Thompson, C.; Chernyashevskyy, O.; Nevirkovets, I.; Metlushko, V.; Rivkin, K.; Ketterson, J.B. (2009). Vortex phase boundaries from ferromagnetic resonance measurements in a patterned disc array. *Phys. Rev. B.* Vol. 80, pp. 014423.

Uhlenbeck, G.E.; Goudsmit,S. (1925). Ersetzung der hypothese vom unmechanischen zwang durch eine forderung bezüglich des inneren verhaltens jedes einzelnen elektrons. *Die Naturwissenschaften.* Vol. 13, Issue (47), pp. 953.

Valenzuela, R.; Zamorano, R.; Alvarez, G.; Gutiérrez, M.P.; Montiel, H. (2007). Magnetoimpedance, ferromagnetic resonance, and low field microwave absorption in amorphous ferromagnets. *J. Non-Cryst. Solids.* Vol. 353, pp. 768.

Valenzuela, R.; Herbst, F.; Ammar, S. (2012). Ferromagnetic resonance in Ni–Zn ferrite nanoparticles in different aggregation states. *J. Magn. Magn. Mater.* Vol. 324, pp. 3398.

Van Vleck, J.H. (1950). Concerning the theory of ferromagnetic resonance absorption. *Phys. Rev.* Vol. 78, pp. 266.

Vargas, J.M.; Zysler, R.D.; Butera, A. (2007). Order–disorder transformation in FePt nanoparticles studied by ferromagnetic resonance. *Appl. Surf. Sci.* Vol. 254, pp. 274.

Vilasi, D. (2001). Hamiltonian dynamics. *World Scientific.* ISBN 981-02-3308-6.

Vittoria, C. (1993). Microwave Properties in Magnetic Films. *World Scientific,* Singapore. pp. 87.

Vlasko-Vlasov, K.; Welp, U.; Jiang, J.S.; Miller, D.J.; Crabtree, G.W.; Bader, S.D. (2001). Field induced biquadratic exchange in hard/soft ferromagnetic bilayers. *Phys. Rev. Lett.* Vol. 86, pp. 4386.

Voges, F.; de Gronckel, H.; Osthöver, C.; Schreiber, R.; Grünberg, P. (1998). Spin valves with CoO as an exchange bias layer. *J. Magn. Magn. Mater.* Vol. 190, pp. 183.

Vounyuk, B.P.; Guslienko, K.Y.; Kozlov, V.I.; Lesnik, N.A.; Mitsek, A.I. (1991). Effect on the interaction of layers on a ferromagnetic resonance in two layer feromagnetic films. *Sov. Phys. Solid State.* Vol. 33, pp. 250.

Walker, L.R. (1957). Magnetostatic modes in ferromagnetic resonance. *Phys. Rev.* Vol. 105, pp. 390.

Wallis, T.M.; Moreland, J.; Riddle, B.; Kabos, P. (2005). Microwave power imaging with ferromagnetic calorimeter probes on bimaterial cantilevers. *J. Magn. Magn. Mater.* Vol. 286, pp. 320.

Wang, X.; Deng, L.J.; Xie, J.L.; Li, D. (2011). Observations of ferromagnetic resonance modes on FeCo-based nanocrystalline alloys. *J. Magn. Magn. Mater.* Vol. 323, pp. 635.

Weil, J.A.; Bolton, J.R.; Wertz, J.E. (1994). Electron Paramagnetic Resonance: elementary Theory and Practical Application. *John Wiley-Sons.* New York. Weil, J.A.; Bolton, J.R. (2007). Electron paramagnetic resonance. *John Wiley & Sons, Inc.* Hoboken, New Jersey. ISBN 978-0471-75496-1.

Weiss, M.T.; Anderson, P.W. (1955). Ferromagnetic resonance in ferroxdure. *Phys. Rev.* Vol. 98, pp. 925.

Wegrowe, J.E.; Kelly, D.; Franck, A.; Gilbert, S.E.; Ansermet, J.Ph. (1999). Magnetoresistance of ferromagnetic nanowires. *Phys. Rev. Lett.* Vol. 82, pp. 3681.

Wegrowe, J.E.; Comment, A.; Jaccard, Y.; Ansermet, J.Ph. (2000). Spin-dependent scattering of a domain wall of controlled size. *Phys. Rev. B.* Vol. 61, pp. 12216.

White, R.L.; Solt, I.H. (1956). Multiple ferromagnetic resonance in ferrite spheres. *Phys. Rev.* Vol. 104, pp. 56.

Wiekhorst, F.; Shevchenko, E.; Weller, H.; Kötzler, J. (2003). Anisotropic superparamagnetism of mono dispersive cobalt-platinum nanocrystals. *Phys. Rev. B.* Vol. 67, pp. 224416.

Wigen P.E.; Zhang, Z. (1992). Ferromagnetic resonance in coupled magnetic multilayer systems. *Braz. J. Phys.* Vol. 22, pp. 267. Wigen, P.E.; Kooi, C.F.; Shanaberger, M.R.; Rosing, T.R. (1962). Dynamic pinning in thin-film spin-wave resonance. *Phys. Rev. Lett.*

Vol. 9, pp. 206. Wigen, P.E. (1984). Microwave properties of magnetic garnet thin films. *Thin Solid Films.* Vol. 114, pp. 135.

Wigen, P.E.; Zhang, Z.; Zhou, L.; Ye, M.; Cowen, J.A. (1993). The dispersion relation in antiparallel coupled ferromagnetic films. *J. Appl. Phys.* Vol. 73, pp. 6338.

Wigen, P.E. (1998). Routes to chaos in ferromagnetic resonance and the return trip: Controlling and synchronizing Chaos, in Bar'yakhtar, V.G.; Wigen, P.E.; Lesnik, N.A. (Eds.). Frontiers in magnetism of reduced dimension systems Nato ASI series (Kluwer, Dordrecht) pp. 29. Wojtowicz, T. (2005). Ferromagnetic resonance study of the free-hole contribution to magnetization and magnetic anisotropy in modulation-doped $Ga_{1-x}Mn_xAs/Ga_{1-y}Al_yAs$: Be. *Phys. Rev. B.* Vol. 71, pp. 035307.

Wolfram, T.; De Wames, R.E. (1971). Magneto-exchange branches and spin-wave resonance in conducting and insulating films: Perpendicular resonance. *Phys. Rev. B.* Vol. 4, pp. 3125.

Woltersdorf, G.; Heinrich, B.; Woltersdorf, J.; Scholz, R. (2004). Spin dynamics in ultrathin film structures with a network of misfit dislocations. *Journal of Applied Physics.* Vol. 95, pp. 7007-7009.

Woods, S.I.; Kirtley, J.R.; Sun, S.; Koch, R.H. (2001). Direct investigation of superparamagnetism in Co nanoparticle films. *Phys. Rev. Lett.* Vol. 87, pp. 137205.

Wüchner, S.; Toussaint, J.C.; Voiron, J. (1997). Magnetic properties of exchange-coupled trilayers of amorphous rare-earth-cobalt alloys. *J. Phys. Rev. B.* Vol. 55, pp. 11576.

Xu, Y.; Zhang, D.; Zhai, Y.; Chen, J.; Long, J.G.; Sang, H.; You, B.; Du, J.; Hu, A.; Lu, M.; Zhai, H.R. (2004). FMR study on magnetic thin and ultrathin Ni-Fe films. *Phys. Stat. Sol. (c).* Vol. 12, pp. 3698.

Yalçın, O.; Yıldız, F.; Özdemir, M.; Aktaş, B.; Köseoğlu, Y.; Bal, M.; Touminen, M.T. (2004(a)). Ferromagnetic resonance studies of Co nanowire arrays. *J. Magn. Magn. Mater.* Vol. 272-276, pp. 1684.

Yalçın, O.; Yıldız, F.; Özdemir, M.; Rameev, B.; Bal, M.; Tuominen, M.T. (2004(b)). FMR Studies of Co Nanowire Arrays, Nanostructures Magnetic Materials and Their Applications. *Kluwer Academic Publisher.* Nato Science Series. Mathematics, Physics and Chemistry. Vol. 143, pp. 347.

Yalçın, O. (2004(c)). PhD Thesis, investigation of phase transition in inorganic spin-Peierls $CuGeO_3$ systems by ESR techgnique. Gebze institute of technology, 2004 Gebze, Kocaeli, Turkey.

Yalçın, O.; Aktaş, B. (2003). The Effects of Zn^{2+} doping on Spin-Peierls transition in $CuGeO_3$ *J. Magn. Magn. Mater.* Vol. 258-259, pp. 137.

Yalçın, O.; Yıldız, F.; Aktaş, B. (2007(a)). Spin-flop and spin-Peierls transition in doped $CuGeO_3$. *Spectrochim. Acta Part A.* Vol. 66, pp. 307.

Yalçın, O. (2007(b)). Comparison effects of different doping on spin-Peierls transition in $CuGeO_3$ *Spectrochim. Acta Part A.* Vol. 68, pp. 1320.

Yalçın, O.; Kazan, S.; Şahingöz, R.; Yildiz, F.; Yerli, Y.; Aktaş, B. (2008(a)). Thickness dependence of magnetic properties of $Co_{90}Fe_{10}$ nanoscale thin films. *J. Nanosci. Nanotech.* Vol. 8, pp. 841.

Yalçın, O.; Erdem, R.; Övünç, S. (2008(b)). Spin-1 model of noninteracting nanoparticles. *Acta Phys. Pol. A.* Vol. 114, pp. 835. Yalçın, O.; Erdem,R.; Demir, Z. (2012). Magnetic properties and size effects of spin-1/2 and spin-1 models of core-surface nanoparticles in different type lattices, smart nanoparticles technology, Abbass Hashim (Ed.), ISBN: 978-953-51-0500-8, InTech, DOI: 10.5772/34706. Available from: http://www.intechopen.com/books/smart-nanoparticles-technology/magnetic-properties-and-size-effects-of-spin-1-2-and-spin-1-models-of-core-shell-nanoparticles-in-di.

Yeh, Y.C.; Jin, J.D.; Li, C.M.; Lue, J.T. (2009). The electric and magnetic properties of Co and Fe films percept from the coexistence of ferromagnetic and microstrip resonance or a T-type microstrip. *Measurement.* Vol. 42, pp. 290.

Yıldız, F.; Yalçın, O.; Özdemir, M.; Aktaş, B.; Köseoğlu, Y.; Jiang, J.S. (2004(a)). Magnetic properties of SmCo/Fe exchange spring magnets. *J. Magn. Magn. Mater.* Vol. 272–276, pp. 1941.

Yıldız, F.; Yalçın,O.; Aktaş, B.; Özdemir, M.; Jiang, J.S. (2004(b)). Ferromagnetic resonance studies on Sm-Co/Fe thin films. *MSMW'04 Symposium Proceedings. Kharkov, Ukraine.*

Yıldız, F.; Kazan, S.; Aktas, B.; Tarapov, S.; Samofalov, V.; Ravlik, A. (2004(c)). Magnetic anisotropy studies on FeNiCo/Ta/FeNiCo three layers film by layer sensitive ferromagnetic resonance technique. *Phys. stat. sol.(c).* Vol. 12, pp. 3694.

Yosida, K. (1957). Magnetic properties of Cu-Mn alloys. *Phys. Rev.* Vol. 106, pp. 893. Yoshikiyo, M.; Namai, A.; Nakajima, M.; Suemoto, T.; Ohkoshi, S. (2012). Anomalous behavior of high-frequency zero-field ferromagnetic resonance in aluminum-substituted ε-Fe2O3. *J. Appl. Phys.* Vol. 111, pp. 07A726.

Young, J.A.; Uehling, E.A. (1953). The tensor formulation of ferromagnetic resonance. *Phys. Rev.* Vol. 93, pp. 544.

Yu, J.T.; Turk, R.A.; Wigen, P.E. (1975). Exchange dominated surface spinwaves in yttrium-iron-garnet films. *Phys. Rev. B.* Vol. 11, pp. 420.

Zakeri, K.; Kebe, T.; Lindner, J.; Farle, M. (2006). Magnetic anisotropy of Fe/GaAs(001) ultrathin films investigated by in situ ferromagnetic resonance. *J. Magn. Magn. Mater.* Vol. 299, pp. L1.

Zakeri, Kh.; Lindner, J.; Barsukov, I.; Meckenstock, R.; Farle, M.; von Hörsten, U.; Wende, H.; Keune, W. (2007). Spin dynamics in ferromagnets: Gilbert damping and two-magnon scattering. *Phys. Rev. B.* Vol. 76, pp. 104416.

Zavoisky, E. (1945). Spin-magnetic resonance in paramagnetics. *J. Phys. USSR.* Vol. 9, pp. 211.

Zianni, X.; Trohidou, K.N. (1998). Monte carlo simulations the coercive behaviour of oxide coated ferromagnetic particles. *J. Phys.: Condens. Matter.* Vol. 10, pp. 7475.

Zhai, Y.; Shi, L.; Zhang, W.; Xu, Y.X.; Lu, M.; Zhai, H.R.; Tang, W.X.; Jin, X.F.; Xu, Y.B.; Bland, J.A.C. (2003). Evolution of magnetic anisotropy in epitaxial Fe films by ferromagnetic resonance. *J. Appl. Phys.* Vol. 93, pp. 7622.

Zhang, Z.; Zhou, L.; Wigen, P.E.; Ounadjela, K. (1994 (a)). Angular dependence of ferromagnetic resonance in exchange coupled Co/Ru/Co trilayer structures, *Phys. Rev. B.* Vol. 50, pp. 6094. Zhang, Z.; Zhou, L.; Wigen, P.E.; Ounadjela, K. (1994 (b)). Using ferromagnetic resonance as a sensitive method to study the temperature dependence of interlayer exchange coupling. *Phys. Rev. Lett.* Vol. 73, pp. 336.

Zhang, B.; Cheng, J.; Gonga, X.; Dong, X.; Liu, X.; Ma, G.; Chang, J. (2008). Facile fabrication of multi-colors high fluorescent/superparamagnetic nanoparticles. *J. Colloid Interf. Sci.* Vol. 322, pp. 485.

Zhu, J.; Katine, J.A.; Rowlands, G.E.; Chen, Y.J.; Duan, Z.; Alzate, J.G.; Upadhyaya, P.; Langer, J.; Amiri, P.K.; Wang, K.L.; Krivorotov, I.N. (2012). Voltage-induced ferromagnetic resonance in magnetic tunnel junctions. *ArXiv: 1205-2835: Cond-mat. Mes-hall.* pp. 1.

Chapter 2

FMR Measurements of Magnetic Nanostructures

Manish Sharma, Sachin Pathak and Monika Sharma

Additional information is available at the end of the chapter

1. Introduction

Ferromagnetic nanowires showed solitary and tunable magnetization properties due to their inherent shape anisotropy. The fabrication of such nanowires in polycarbonate track-etched and anodic alumina membranes have been widely studied during the last 15 years [1-2]. Their potential applications might be explored in spintronic devices and more specifically in magnetic random access memory (MRAM) and magnetic logic devices [3-5]. Furthermore, microwave devices, such as circulators or filters for wireless communication and automotive systems can be fabricated on ferromagnetic nanowires embedded in AAO substrates [6-9].

This chapter begins with a brief overview of the historical development of the theory of ferromagnetic resonance in magnetic nanostructures. State-of-the-art calculations for resonance frequency in ferromagnetic nanowires (solid and hollow) and multilayer nanowires are presented. In addition, experimental approach to synthesis such structures and detecting material properties using various techniques will be discussed in brief. Recently, due to the development of spintronics, there have been increasing interests in the microwave dynamics of one-dimensional structures such as nanowires and two dimensional structures like multilayer magnetic films. The most important parameters that control dynamic behaviors are the internal fields and damping constant. The ferromagnetic nanowires in anodic alumina (AAO) templates seem to be attractive substrates for microwave applications. Since they have high aspect ratio, electromagnetic waves can easily penetrate through them. They exhibit ferromagnetic resonance (FMR) even at zero bias fields and, due to their high saturation magnetization, operating frequency can be tuned with DC fields.

FMR is a useful technique in the measurement of magnetic properties of ferromagnetic materials. It has been applied to a range of materials from bulk ferromagnetic materials to nano-scale magnetic thin films and now a day's people have started research to characterise nanoparticles and nanowires systems. The dynamic properties of magnetic materials can be easily perturbed by ferromagnetic resonance (FMR), as they can excite standing spin waves due to magnetic pinning [10-11,31,32]. It also yields direct information about the uniform precession mode of the nanowires which can be related to the average anisotropy magnitude [12-15]. Several measurement techniques which have been used to characterise magnetization dynamics such as femtosecond spectroscopy [16-18], pulse inductive microwave magnetometer [19], FMR force microscopy [20], network analyzer FMR [21,31,32] and high-frequency electrical measurements of magnetodynamics [22]. All these techniques can be used for modern application for example in telecommunications and data storage systems.

In the present chapter, we deal with magnetic nanostructures such as nanowires and multilayered nanodiscs and rings which exhibit unique FMR responses since their various anisotropy energies are strongly influenced by size and shape [33]. In addition, interactions between multilayered segments can be tailored such that the FMR response is not only angle-dependent but also influenced strongly at certain frequencies [32]. In this chapter, we shall be describing three different aspects of this topic:

1.   We first develop the theory of FMR response of densely packed nanowire arrays that can be treated as two-dimensional periodic nanostructures. We then extend the theory to three-dimensional structures, which can be made using multilayered nanowires.
2.   We then describe how to synthesize such nanowire arrays and also direct measurements of such structures. Several different experimental techniques are discussed.
3.   We then continue the treatment to describe the use of such periodic nanowire arrays in microwave devices to exhibit nonlinear responses and also for circulators and isolators. Although the effects seen till now are weak, these are still quite promising.

2. Fundamental theory of ferromagnetic resonance

Ferromagnetic resonance (FMR) is a very powerful experimental technique in the study of ferromagnetic nanomaterials. The precessional motion of a magnetization M of ferromagnetic material about the applied external magnetic field H is known as the Ferromagnetic resonance (FMR). In the physical process of resonance, the energy is absorbed from rf transverse magnetic field h_{rf}, which occurred when frequency matched with precessional frequency (ω). The precession frequency depends on the orientation of the material and the strength of the magnetic field. It allows us to measure all the most important parameters of the material: Curie temperature, total magnetic moment, relaxation mechanism, elementary excitations and others.

A single domain magnetic particle with ellipsoid shape was considered as in [Kittel] to drive the resonance condition for the phenomenon of ferromagnetic resonance. A uniform, static magnetic field H is applied along the z-axis and set the sample in a microwave cavity. A resonance is observed at a frequency given by

$$\hbar\omega = g\mu_B \sqrt{[H + (N_x - N_z)M][H + (N_y - N_z)M]}. \tag{1}$$

N_x, N_y, N_z are the demagnetization factors in the x, y, z directions, and so on, and where g is the spectroscopic splitting factor (Lande factor) and $\mu_B = eh/2m_e$ is the Bohr magneton. The demagnetizing factors affect the shape anisotropy of the magnetic material depending upon its geometry to be ellipsoid, sphere, thin film etc.

Part of the classical approach to ferromagnetism is to replace the spins by a classical micro-spin vector M magnetization. The time-dependence of the magnetization can be obtained directly by calculating the torque acting on M by an effective field H_{eff},

$$\frac{dM}{dt} = -\gamma M \times H,$$ (2)

where $\gamma = g\mu_b/\hbar$ is a gyromagnetic ratio. This equation represents an undamped precession of the magnetization. From experiments actual changes of the magnetization are known to decay in a finite time. The occurrence of a damping mechanism leads to reversal of the magnetization towards the direction of H within several nanoseconds. The damping is just added as a phenomenological term to Eq. 2.

$$\frac{dM}{dt} = -\gamma M \times \left(H_{eff} - \frac{\alpha}{\gamma M_s} \frac{dM}{dt} \right) = -\gamma M \times H_{eff} + \frac{\alpha}{M_S} M \times \frac{dM}{dt}$$ (3)

where α is the dimensionless Gilbert damping constant, of order 10^{-2} in ferromagnetic thin films. Eq. 3 is known as Landau-Lifshitz-Gilbert equation after Gilbert introduces the damping term. H_{eff} is the total effective magnetic field which is a sum of static applied magnetic field (H), dynamic magnetic field (h_{rf}) and internal magnetic field (H_{in}). Internal field constitutes various magnetic anisotropies such as magnetocrystalline anisotropy, shape anisotropy, and magnetoelastic anisotropy etc. This equation therefore describes torque acting on M. This torque leads to a rotation of the magnetization towards the direction of the external magnetic field. The damping causes decay in precessional motion which by applying a dynamic magnetic field becomes continuous as shown in Fig. 1.

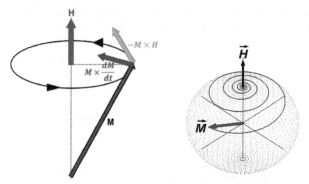

Figure 1. (a) Torque components exerted on the magnetization M by rotational field H (b) Motion of M for constant H.

3. FMR dispersion relation for nanostructures

Due to miniaturization from bulk to the nanoscale, material properties shows a drastic change like surface-to-volume ratio, electron transport, thermodynamic fluctuations, defects etc. The nanomagnetic material demonstrates distinct magnetic response due to various anisotropies. The one-dimensional nanostructures such as nanowires, nanotubes, nanorods and nanorings are the area of recent research for data storage applications, sensors, biomedical drugs, microwave devices. To perturb the dynamic magnetization in these structures, ferromagnetic resonance is an effective tool.

In case of a system which incorporates an array of nanowires dipole-dipole interaction, shape anisotropy, crystalline anisotropy and Zeeman energy interaction plays a complex role. Therefore the total energy will be the sum of all internal energies.

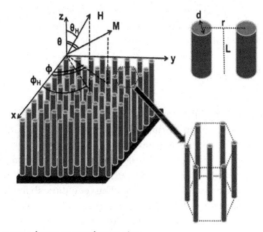

Figure 2. Coordinate system for an array of nanowires

Fig. 2 shows the schematic of an array of nanowires with relative orientation of the magnetization M and the applied magnetic field H *w.r.t* nanowire axis in spherical coordinate systems. The free energy density equation for an array of magnetic nanowires in the presence of external magnetic field at angles (θ_H) from nanowires axis can be written as

$$E \approx -MH\left(\sin\theta\sin\theta_H\cos(\phi-\phi_H)+\cos\theta\cos\theta_H\right)+K_{eff}\sin^2\theta \tag{4}$$

where K_{eff} is the effective uniaxial anisotropy which can be written as

$$K_{eff} = \pi M^2\left(1-3P\right)+K_U \tag{5}$$

The first term includes the dipole-dipole interaction between the nanowires and second term represents second-order uniaxial anisotropy along the wire axis. P is the porosity which can be obtained from

$$P = \frac{\pi}{2\sqrt{3}} \frac{d^2}{r^2} \tag{6}$$

where d is the diameter of the pore and r is the centre-to-centre inter-wire distance between the pores. The equilibrium values for polar angles are obtained by minimizing the energy term

$$E_\theta = \frac{\partial E}{\partial \theta} = 0 \tag{7}$$

$$E_\theta = -MH\left(\cos\theta\sin\theta_H \cos(\phi - \phi_H) - \sin\theta\cos\theta_H\right) + K_{eff}\sin 2\theta = 0 \tag{8}$$

From eq (8) we retrieve the dispersion relation which can be written as

$$\left(\frac{\omega}{\gamma}\right)^2 = \left[\text{Hcos}(\theta - \theta_H) + \text{H}_{eff}\cos^2\theta\right]\left[\text{Hcos}(\theta - \theta_H) + \text{H}_{eff}\cos 2\theta\right] \tag{9}$$

Here, $H_{eff} \approx 2K_{eff}/M_S$ is the effective anisotropy field that comes from a combination of effects including shape, magnetocrystalline and magnetoelastic anisotropy. Therefore from eq (5)

$$H_{eff} = 2\pi M_S(1 - 3P) + 2\frac{K_U}{M_S} \tag{10}$$

For the case of multilayer nanowires, the effective anisotropy field H_{eff} is given by

$$H_{eff} = 2\pi M_S\left(1 - 3\{1 - f(1 - P)\}\right) + 2\frac{K_U}{M_S} \tag{10a}$$

where, $f \approx \frac{h_m}{h_{nm} + h_m}$, h_m and h_{nm} represents magnetic and non-magnetic thicknesses in multilayer section. If $h_{nm} = 0$ (i.e. no non-magnetic spacer layer), the above equation represents the case of a single-element nanowire. Depending upon the direction of the external magnetic field along the easy axis of the nanowires, we can determine the various cases:

Case 1: H∥ to the wire and $H_{eff} > 0$

$$\frac{\omega}{\gamma} = H + H_{eff} \tag{11}$$

Case 2: H∥ to the wire and $H_{eff} < 0$

$$\frac{\omega}{\gamma} = H_{eff} - H \tag{12}$$

Case 3: $H_\perp$ easy axis and $H < H_{eff}$

$$\left(\frac{\omega}{\gamma}\right)^2 = \left(H^2_{eff} - H^2\right) \tag{13}$$

Case 4: $H_\perp$ easy axis, $H > H_{eff} > 0$

$$\left(\frac{\omega}{\gamma}\right)^2 = H(H - H_{eff}) \tag{14}$$

The frequency-field characteristics can be studied from these relations for various cases of the direction of the applied field and corresponding angular variation with resonance field of the nanowires as shown in Fig. 3. The horizontal line shows the intersection of the dispersion relation and indicates where in an FMR spectrum the resonance lines would be found for a fixed frequency in Fig. 3(a).

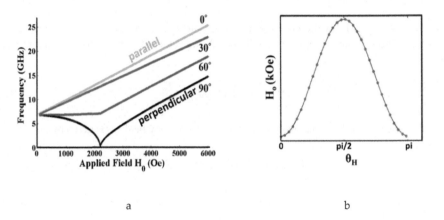

a   b

Figure 3. Simulated plots (a) dispersion relation for Ni nanowires for various angles in which the external magnetic field is applied (b) resonance field as a function of the externally applied field angle θ_H.

4. Sample preparation and characterization

During last decade magnetic nanowires have attracted enormous research attention in many areas of advanced nanotechnology, including patterned magnetic recording media, materials for optical and microwave applications. Template assisted growth of nanostructures under constant potential in several electrolytes has been carried out by researchers for over 30 years. In comparison to other deposition techniques (sputtering and MBE) electrodeposition is a low cost and simple technique to fabricate magnetic nanowires and multilayers. Arrays of Co nanowires were fabricated by template assisted electrochemical deposition into the nanometer-sized pores.

5. Growth of nanostructures

Templates

Nanowires were grown by using two different types of templates (commercially available) polycarbonate track etched (PCTE) and aluminum oxide (AAO) from Whatman. These templates are completely different from each other due to their qualities as well as the preparation methods. The methods used for fabrication and properties of templates are given below;

Polycarbonate track etched (PCTE)

In PCTE template, the pore size varies from 20nm to 200nm and the thickness of ~6μm. To fabricate the PCTE template, initially high energy particle are used to bombard it to produce the path. Later these paths are etched in different chemical bath. The size of the pores is determined by the etching process. Fig.4(a) shows a SEM picture of a commercial PCTE membrane, with a reported pore size of 100 nm. Although the pores seem to have similar diameters, the pore placement is random. The pores density is quite low (in the range of 10^{10} to 10^{12} per cm²) in case of PCTE.

Anodic aluminum oxide (AAO)

In 1995, Masuda and Fukuda reported the method to fabricate highly ordered nanohole arrays on aluminum foil. Double anodization of aluminium in acidic solution is adopted to fabricate porous alumina template. The beauty of these templates is their cylindrical pores of uniform diameter, arranged in hexagonal arrays with a thin oxide layer exist at the bottom. Anodic Aluminum Oxide (AAO) templates have been successfully used as templates for the growth of nanowires by electrodeposition. In order to fabricate the AAO template the aluminum is cleaned in the acidic medium to remove the surface impurities. Further the cleaned aluminium processed by two step anodization described by Masuda. The pore size and spacing between pore in the alumina templates are controlled by the anodization voltage.

Temperature of electrolyte also plays a important role in pore parameters. The pores in AAO template cylindrical and highly dense which makes it a good candidate to study the interaction effect between the nanowires on magnetic properties. Fig. 4(b)-(c) shows the SEM micrograph of AAO templates.

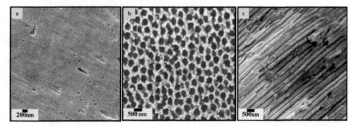

Figure 4. SEM micrograph of membranes (a) Polycarbonate (b) Anodic alumina (Top view) and (c) anodic alumina (cross-section view).

Template assisted electrodeposition technique

Electrodeposition process involves the electric current to reduce cations from electrolyte and deposited that material as a thin film onto a conducting substrate. At the cathode the metal reduction takes place and metal deposits according to:

$$Mn^+ + n\ e^- = M(s)$$

In order to form the nanowires, the cations from electrolyte move through the non-conducting template (AAO/ PCTE) having nanosized pores and deposited on the conducting substrate.

The desired material properties depends upon the various process parameters like electrolyte composition, bath pH, mode of deposition (DC, pulse and AC) and deposition temperature.

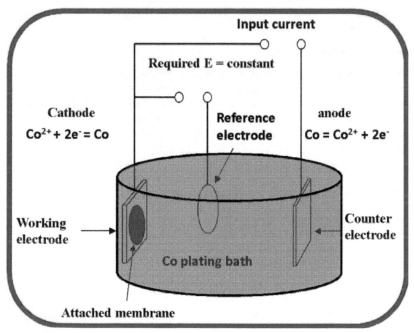

Figure 5. Schematic representation of three-electrode electrochemical cell setup employed. AAO template mounted electrodes act as a working electrodes (WE), platinum foil counter electrode (CE) and Saturated calomel electrode (SCE), reference electrode (RE).

Fig.5 illustrates the three-electrode cell set-up used in this study. A platinum foil and a saturated calomel electrode (SCE) were used as the anode (or counter electrode) and as a reference electrode respectively. The steps of preparation of nanostructure are as followed;

Steps involved in the template assisted synthesis of nanowires

Step 1: The porous anodic alumina (AAO) /polycarbonate templates were taken and one side of the AAO were sputtered with Au by RF sputtering, which acted as the working electrode in a three-electrode electrochemical cell.

Step 2: The electrodeposition solution was restrained to the other side of the membrane so that deposition was initiated onto the Au layer within the pores. The array of Co nanowires was deposited from a solution of 25 gm/L $CoSO_4.7H_2O$, 5 gm/L H_3BO_3 and sodium Lauryl Sulphate (SLS), which is used to reduce the surface tension of water for proper wetting of pores.

Step 3: The cyclic voltammetry is used to figure out the constant deposition potential of the working electrode with respect to a standard reference electrode **RE** to get the favourable condition for deposition.

Step 4: The growth of nanowires is carried at the optimized potential.

Step 5: For further investigation of freely standing nanowires, the templates is used to dissolve in an appropriate solution. Etching solution for AAO and PCTE are sodium hydro-oxide and dichloromethane respectively.

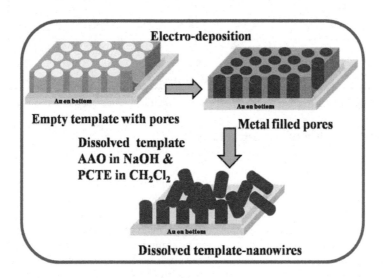

Figure 6. Schematic illustration of the growth of magnetic nanowire in alumina template by electrodeposition process.

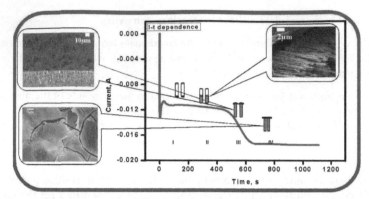

Figure 7. Typical choronoamperometery plot during potentiostatic electrodeposition taken during the fabrication of Co nanowires. The various stage of pore filling during deposition is shown as insets at the respective current-time positions.

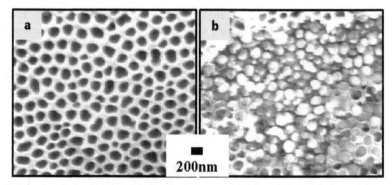

Figure 8. Scanning electron micrograph (SEM) of empty and filled surface morphology of the template.

6. Characterization of nanostructures

Scanning electron microscopy

For scanning electron microscopy (SEM), the template containing nanowires is partially released from their template by appropriate solution. To remove the residual part of template the etched sample is cleaned by deionised water. To carry out the surface analysis secondary electron imaging in scanning electron microscope has been utilized. Surface morphology of Co nanowire was investigated by scanning electron microscope (SEM: ZEISS EVO 50) operating at 20 kV accelerating voltage by secondary electron imaging. Fig. 8 shows the SEM micrograph of empty AAO and filled with nanowires. The morphology of grown nanowires in PCTE template can be clearly seen in Fig. 9(a). Side and top view of AAO assisted nanowires are presented in Fig. 9(b & c). The growth density of nanowires in AAO template is more as compared to PCTE template.

Figure 9. Scanning electron micrograph (SEM) of electrodeposited nanostructure in (a) Polycarbonate and (b) & (c) Anodic alumina.

Transmission electron microscopy

Transmission Electron Microscope (TEM) is a widely used instrument for characterizing the interior structure of materials. For TEM, the template was completely etched and rinsed several times with deionised water to clean the residual template part. Template etching is an important step of sample preparation of nanowires for TEM characterization. Cobalt nanowires were scratched from the substrate and ultrasonicated in acetone for 15 min so that the nanowires could disperse properly. Few drops of the suspension were then transferred on to a carbon coated copper grid and the microstructures were analyzed by high resolution TEM (HRTEM: Technai G20 S-Twin model) operating at 200 kV. To get the actual diameter of the nanowire, the complete dissolving of residual template from surrounding the nanowires is very important. Typical TEM results obtained are shown in Figure 10. While nanowires grown in PCTE templates typically have a tapered cross-section due to the non-uniform diameters of the pores, the AAO template-based nanowires grow in a more uniform manner.

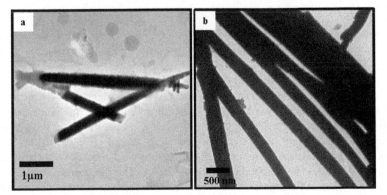

Figure 10. Transmission electron micrograph (TEM) of electrodeposited nanostructure in (a) Polycarbonate and (b) Anodic alumina.

7. Electron paramagnetic resonance

EPR/FMR measurements obtained from ferromagnetic nanowire arrays give detailed information on the size of the nanowires. The spectra can be used to calculate the interwire magnetic interactions quite accurately [31]. In Fig. 11 are shown typical spectra obtained from Co nanowire arrays. Here, both single-Co nanowire arrays and multilayered Co/Pd nanowire arrays are compared. There is clearly an angular dependence of the applied microwave field with respect to the easy axis of the nanowires, as has been studied previously [32].

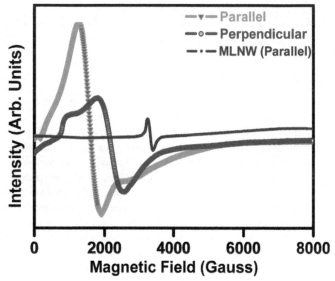

Figure 11. Electron paramagnetic resonance (EPR/FMR) spectra of single element Co nanowire arrays in parallel and perpendicular orientation at room temperature.

Fig.12 shows the FMR spectra of Co/Pd MLNW arrays for the temperature range from 90K to 290K. It is clear from the data that two different peaks in the FMR spectrum have been observed. The peak present at the ~ 3310 G is corresponding to the main peak of FMR and it is due to the quantum confinement in the nanostructures which is more in case of Co/Pd multilayered nanowires. The spin waves are confined in the magnetic nanostructure and it is found to be dominant when we reduce the third dimension (in z-direction) of the nanostructure. This main FMR peak present at all the temperature and the variation of resonance field at different temperature is very slow. As we go down to the 90K with decrease of 20K from room temperature the resonance field increases slowly. This slow variation can be explain as temperature decreases the M_s (T) increase and resultant head-to-tail alignment between FM segments increases, which further reduces the effective field as a result slow increase in the resonance field. Inset in Fig.12 shows the variation of resonance field with temperature.

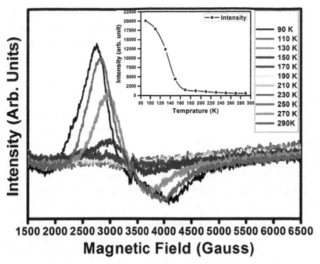

Figure 12. Electron Spin resonance (ESR/FMR) spectra of multilayered Co/Pd nanowire arrays in parallel orientation with temperature range (90K-290K). (Inset) Peak height change with temperature.

8. Microwave devices

The increasing demand for higher frequency magnetic microwave structures triggered a tremendous development in the field of magnetization dynamics over the past decade. In order to develop smaller and faster devices, a better understanding of the complex magnetization precessional dynamics, the magnetization anisotropy, and the sources of spin scattering at the nanoscale is necessary [23-25]. Magnetic data storage with its promise of non-volatility, robustness, high speed and low energy dissipation attracted long back. It has been encountered in a number of applications such as smart cards, hard drives, thin film read heads or video tapes [26-28]. Industry activities are also directed to replace semiconductor random access memories by magnet based memory devices. Logical 0 and 1 are encoded by the direction of the magnetization of a small magnetic element, such as a giant magnetoresistance (GMR) or tunneling magnetoresistance (TMR) element. For data manipulation *i.e.* reading and writing, the magnetization has to be switched between the two equilibrium positions. In order to push forward data transfer rates which are already in the GHz range, spin waves transportation mechanism has to be employed in place of current semiconductor based devices [29-32]. To this end an understanding of the dynamic motion and the mode spectrum is necessary.

On the road towards the size reduction of microwave devices, ferromagnetic nanowires embedded into porous templates have proven to be an interesting route to ferrite based materials. The main advantages of what we call ferromagnetic nanowired (FMNW) substrates are that they present a zero-field microwave absorption frequency that can be easily tuned over a large range of frequencies, as well as being low cost and fast to produce over a large area as compared to standard ferrite devices. Conventional ferrite circulators need to be biased by a magnetic field to operate. This biasing field is generally provided by permanent magnet and in view of the volume reduction, we need to think of FMNW substrates which work at zero fields. A rigorous theory of microwave devices is very cumbersome due to ferromagnetic resonance (FMR) and spin-wave phenomena.

VNA-FMR is an important technique for the investigation of magnetization dynamics of low-dimensional magnetic structures and patterned microwave devices. Also, broadband flip-chip technique can be used to measure the material intrinsic and extrinsic properties by applying external magnetic field. In this chapter, we will demonstrate both techniques to have better understanding of ferromagnetic resonance. We propose fabrication and measurements of various non reciprocal microwave devices like band-stop filter, isolator and electromagnetic band-gap (EBG) structures using FMNW substrates using Ni and Co nanowires.

9. Conventional Ferromagnetic Resonance – Flip-chip based technique (FMR)

In the frequency domain measurements, the magnetic excitation is sinusoidal magnetic field h_{rf} and the response of the sample is detected by vector network analyzer. The magnetic field can be applied along parallel and perpendicular direction of the nanowires satisfying the FMR condition. The main components of the experimental setup are shown in Fig. 13. The VNA is connected to a coplanar waveguide (CPW) having a characteristic impedance of 50 Ω using coaxial cables and microwave connectors. For such radio frequency connections often coaxial cables with Teflon insulation and SMA connectors are employed. They are comparably low priced and offer a bandwidth of typically 18 GHz. The used cables should not have a metallic reinforcement.

Flip-chip based measurements are done to extract the material parameters, which are technologically important for data storage applications [31]. Fig. 13 shows the schematic of the set-up used for flip-chip measurements. Magnetic field is applied parallel to the nanowires and perpendicular to RF magnetic field, so that resonance condition must be satisfied.

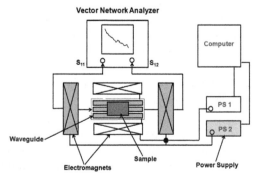

Figure 13. Schematic representation of VNA-FMR system

The transmission response of Ni NWs placed on a co-planar waveguide transmission line (Fig. 14(a)) for different static magnetic field is shown in Fig. 14(b). A small but perceptible and repeatable change was observed when the applied magnetic field was turned on/off. The change is however too small to show up in the plot of Fig. 14(b). This is partly due to masking of the shifts due to reflections from soldering and other undesirable metal deposits near the patterned transmission line and partly due to the relatively small interaction between the RF fields (largely confined to the substrate holding the CPW line) and the nanowires. It is thus expected that clearer shifts can be seen with more careful preparation and patterning of the substrate and ultimately by printing the CPW line on the nanowire template itself.

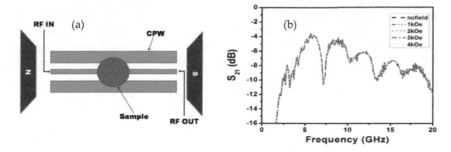

Figure 14. (a) CPW geometery for magnetic field applied along perpendicular direction (b) transmission response as a function of frequency of Ni nanowires for various applied magnetic field

10. Substrate integrated magnetic microwave devices

Newly developed techniques enable to characterize and re-arrange matter at nanometer scale. Now a day, automotive and wireless communication requires reduction in dimensions of nanodevices for yielding higher cut-off frequencies. Here, we propose integrated

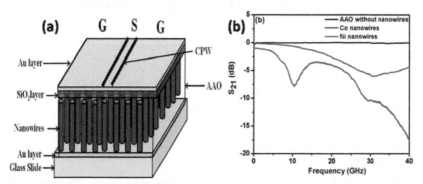

Figure 15. schematic of band stop filter (b) S-parameter of the DUT for three different samples

magnetic band-stop filter fabricated on FMNWS and studying their microwave properties like permittivity and permeability using CPW. Fig. 15(a) shows the schematic of the device under test which is a coplanar waveguide on NW substrate. The transmission coefficient of the device for three samples having bare AAO, Ni and Co NWs are shown in Fig. 15(b) at zero biasing. It is observed that by applying the magnetic field the resonance frequency shift towards higher range, so that we can tune our operating frequency of the device. We also observe that material properties also influence the Device-Under-Test (DUT).

11. Non-reciprocal devices

It is important to note that it is possible to obtain non-reciprocal structures also using nanowire-based devices. Essentially, the microwave properties like permittivity and permeability can be made to be asymmetric. This is useful to obtain non reciprocal devices such as isolators and circulators. Prototypes of such devices are presently under fabrication. Some of the proposed devices are shown in Fig. 16 below. The detailed studies of these devices are out of scope of this chapter.

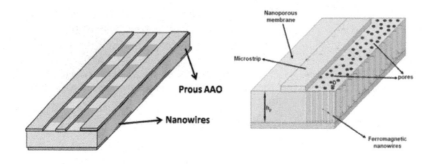

EBG structure in porous AAO template   Isolator

Figure 16. Possible non-reciprocal structures using nanowires.

Author details

Manish Sharma, Sachin Pathak and Monika Sharma
Indian Institute of Technology Delhi, India

12. References

[1]   A. Fert and L. Piraux, "Magnetic nanowires", J. Magn. Magn. Mater. 200 (1999) 338-358.

[2] S. Shamaila, R. Sharif, S. Riaz, M. Khaleeq-ur-Rahman and X. F. Han, "Fabrication and magnetic characterization of Co_xPt_{1-x} nanowire arrays", Appl. Phys. A, 92 (2008) 687-691.

[3] Z. Z. Sun and J. Schliemann, "Fast domain wall propagation under an optimal field pulse in magnetic nanowires", Phys. Rev. Lett. 104 (2010) 037206.

[4] M. Yan, A. Kakay, S. Gliga and R. Hertel, "Beating the walker limit with massless domain walls in cylindrical nanowires", Phys. Rev. Lett. 104 (2010) 057201.

[5] C. T. Boone, J. A. Katine, M. Carey, J. R. Childress, X. Cheng and I. N. Krivorotov, "Rapid domain wall motion in permalloy nanowires excited by a spin-polarized current applied perpendicular to the nanowire", Phys. Rev. Lett. 104 (2010) 097203.

[6] B. K. Kuanr, V. Veerakumar, R. Marson, S. Mishra, R. E. Camley and Z. Celinski, "Nonreciprocal microwave devices based on magnetic nanowires", Appl. Phys. Lett. 94 (2009) 202505.

[7] M. Darques, J. De La T. Medina, L. Piraux, L. Cagnon and I. Huynen, "Microwave circulator based on ferromagnetic nanowires in alumina template", Nanotechnology 21 (2010) 145208.

[8] J. De La T. Medina, J. Spiegel, M. Darques, L. Piraux and I. Huynen, "Differential phase shift in nonreciprocal microstrip lines on magnetic nanowired substrates", Appl. Phys. Lett. 96 (2010) 072508.

[9] C. Kittel, "On the theory of ferromagnetic resonance absorption", Phys. Rev. 73 (1948) 155.

[10] C. Kittel, "Excitation of spin waves in a ferromagnet by a uniform rf field", Phys. Rev. 110 (1958) 1295-1297.

[11] L. Kraus, G. Infante, Z. Frait and M. Vazquez, "Ferromagnetic resonance in microwires and nanowires", Phys. Rev. B 83 (2011) 174438.

[12] J. De La T. Medina, L. Piraux, J. M. Olais Govea and A. Encinas, "Double ferromagnetic resonanace and configuration-dependent dipolar coupling in unsaturated arrays of bistable magnetic nanowires", Phys. Rev. B 81 (2010) 144411.

[13] A. Encinas-Oropesa, M. Demand, L. Piraux, I. Huynen and U. Ebels, "Dipolar interactions in arrays of nickel nanowires studied by ferromagnetic resonance", Phys. Rev. B 63 (2001) 104415.

[14] C. A. Ramos, M. Vazquez, K. Nielsch, K. Pirota, R. B. Rivas, R. B. Wherspohn, M. Tovar, R. D. Sanchez and U. Gosele, "FMR characterization of hexagonal arrays of Ni nanowires", J. Magn. Magn. Mater. 272-276 (2004) 1652-1653.

[15] C. A. Ramos, E. Vassallo Brigneti and M. Vazquez, "Self-organized nanowires: evidence of dipolar interactions from ferromagnetic resonance measurements", Physica B 354 (2004) 195-197.

[16] E. Beaurepaire, J. C. Merle, A. Daunois and J. Y. Bigot, "Ultrafast spin dynamics in ferromagnetic nickel", Phys. Rev. Lett. 76 (1996) 4250-4253.

[17] M. R. Freeman, and W. K. Hiebert, "Stroboscopic microscopy of magnetic dynamics", In: B. Hillebrands, K. Ounadjela (Eds.), Spin Dynamics in Confined Magnetic Structures I. Springer, Berlin, pp. 93-126.

[18] W. K. Hiebert, A. Stankiewicz, and M. R. Freeman, "Direct observation of magnetic relaxation in a small permalloy disk by time-resolved scanning Kerr microscopy", Phys. Rev. Lett. 79 (1997) 1134-1137.

[19] T. J. Silva, C. S. Lee, T. M. Crawford, and C. T. Rogers, "Inductive measurement of ultrafast magnetization dynamics in thin-film permalloy", J. Appl. Phys. 85 (1999) 7849-7862.

[20] M. M. Midzor, P. E. Wigen, D. Pelekhov, W. Chen, P. C. Hammel, and M. L. Roukes, "Imaging mechanisms of force detected FMR microscopy", J. App. Phys. 87 (2000) 6493-6495.

[21] O. Mosendz, B. Kardasz, D. S. Schmool, and B. Heinrich, "Spin dynamics at low microwave frequencies in crystalline Fe ultrathin film double layers using co-planar transmission lines", J. Magn. Magn. Mater. 300 (2006) 174-178.

[22] N. Mecking, Y. S. Gui, and C. M. Hu, "Microwave photovoltage and photoresistance effects in ferromagnetic microstrips", Phys. Rev. B 76 (2007) 224430.

[23] B. K. Kuanr, M. Buchmeier, D. E. Buergler, P. Gruenberg, R. E. Camley, and Z, Celinski, "Dynamic and static measurements on epitaxial Fe-Si-Fe", J. Vac. Sci. Technol. A 24 (2003) 1157.

[24] J. Curiale, R. D. Sanchez, C. A. Ramos, A. G. Leyva, A. Butera, "Dynamic response of magnetic nanoparticles arranged in a tubular shape", J. Magn. Magn. Mater. 320 (2008) 218-221.

[25] Can-Ming Hu, "Recent progress in spin dynamics research in Canada", La Physique Au Canada 65 (2009).

[26] X. Kou, X. Fan, R. K. Dumas, Q. Lu, Y. Zhang, H. Zhu, X. Zhang, K. Liu and J. Q. Xiao, "Magnetic effects in magnetic nanowire arrays", Adv. Mater. 23 (2011) 1393-1397.

[27] T. Schrefl, J. Fidler, D. Suss, and W. Scholz, "Hysteresis and switching dynamics of patterned magnetic elements", Phys. B 275 (2000) 55-58.

[28] X. Kou, X. Fan, H. Zhu and J. Q. Xiao, "Ferromagnetic resonance and memory effects in magnetic composite materials".

[29] K. Nagai, Y. Cao, T. Tanaka and K. Matsuyama, "Binary data coding with domain wall for spin wave based logic devices", J. Appl. Phys. 111 (2012) 07D1301-07D1304.

[30] M. Sharma, S. Pathak, S. Singh, M. Sharma and A. Basu, "Highly ordered magnetic nickel nanowires: structural properties and ferromagnetic resonance measurements", AIP Conf. Proc. 1347 (2011).

[31] O. Yalçın, et al, Ferromagnetic resonance studies of Co nanowire arrays, Journal of Magnetism and Magnetic Materials 272–276 (2004) 1684–1685.

[32] G. Kartopu, O. Yalçın, et al. Size effects and origin of easy-axis in nickel nanowire arrays, Journal of Applied Physics 109, 033909 (2011).

[33] G. Kartopu, O. Yalçın, M. Es-Souni, and A. C. Başaran, Magnetization behavior of ordered and high density Co nanowire arrays with varying aspect ratio, Journal of Applied Physics 103, 093915 (2008).

Chapter 3

FMR Studies of [SnO$_2$/Cu-Zn Ferrite] Multilayers

R. Singh and S. Saipriya

Additional information is available at the end of the chapter

1. Introduction

The artificially structured multilayers (ML) have opened a new field of interface magnetism. It is possible to fabricate a multilayer sample with a specific design according to the requirement consisting of magnetic, non-magnetic, metallic or non-metallic components. Various types of interfaces may be synthesized in ML by combining a magnetic component with a non-magnetic one or from two different magnetic components. The interface between two non-magnetic components is particularly interesting if a magnetic anomaly happens to be induced at the interface atom layer due to some interface effect.

The oscillations of magnetic parameters as a function of the number of interfaces, spacer layer or magnetic layer thickness in the ML have been gaining attention in the recent years. Bruno et al [1] employed RKKY model and Edwards et al [2] used spin dependent confinement of electrons in the quantum well provided by the spacer layer (quantum interference) to explain the oscillations initially observed in metallic ML. Exchange coupling in ML is also affected by direct dipolar coupling like the correlation of spins at rough interfaces (orange peel coupling), which is magnetostatic in origin and occurs due to the interaction of the dipoles which appear due to the roughness of the material [3]. This coupling favors parallel or anti-parallel alignment of spins depending upon the interplay between the magnetostatic exchange and anisotropy energy.

The oscillations in exchange coupling are also observed in oxide ML apart from the metallic ML [4]. Oxide materials are more stable and their ML can form magnetic structures which do not exist in bulk form. There are reports on ML of ferrites used along with NiO [5,6], MgO [7], CoO [8] and SnO$_2$[9-12].

The mixed spinel structure of Cu-Zn ferrites with composition $Cu_{0.6}Zn_{0.4}Fe_3O_4$ (CZF) gives high magnetization when deposited in Ar environment [13]. The properties of non-magnetic (NM) SnO$_2$ are sensitive to oxygen vacancies [14]. The interfacial region between these two materials may give rise to interesting magnetic phenomena. The objective of the present work is to study the magnetic properties of SnO_2 and Cu-Zn Ferrite multilayers and interpret them in terms of suitable magnetization mechanism.

This chapter is about the Ferromagnetic Resonance (FMR) studies of (SnO_2/Cu-Zn Ferrrite) ML as a function of CZF layer thickness and SnO_2 layer thickness.

2. Experimental

Alternate layers of SnO_2 and $Cu_{0.6}Zn_{0.4}Fe_3O_4$ (CZF) were deposited on quartz substrates at room temperature (RT) using rf-magnetron sputtering in Ar environment from SnO_2 and CZF targets at a power of 50W and 70W respectively. The rate of deposition was estimated by depositing thin films of SnO_2 and CZF separately. The synthesis method of targets and the rf sputtering system are described in our earlier work [15]. The field emission scanning electron microscopy (FESEM) studies were carried out to observe the multilayer structure. The FMR studies were carried out in the temperature range 100 - 450 K using JEOL x-band spectrometer. FMR studies were carried out at different temperatures and by varying the angle θ_H between the thin film normal and the direction of the applied field. The peak-to-peak FMR signal intensity (I_{PP}) which is proportional to the power absorbed was measured between the positive and negative peak points along the y-axis. The peak-to-peak linewidth (ΔH) was measured as a function of temperature.

3. Ferromagnetic Resonance (FMR) studies

3.1. Effect of CZF layer thickness

The ML samples [SnO_2 (46 nm)/CFZ(x nm)]$_5$ where x = 42 nm, 83 nm and 249 nm were synthesized. A final capping layer of 46 nm of SnO_2 was deposited on all the ML samples. The error in the thickness is ± 5 nm. The FESEM images (figure 1) portray a stack of alternate dark and light layers corresponding to CZF and SnO_2 respectively as confirmed by the EDAX data. The CZF layer exhibits a columnar growth as observed in many of the sputtered films [16]. The protrusions of CZF column into SnO_2 layers visible from the FESEM images indicate that the interfaces are diffused.

FMR studies were carried out at different temperatures and by varying the angle θ_H between the thin film normal and the direction of the applied field. Figure 2 shows the room temperature FMR spectra of the ML in θ_H = 90° (parallel) and 0° (perpendicular) configuration. In the parallel geometry, the FMR signal for the ML with x = 249 nm is highly asymmetric. The steep increase in the low field side and a slow increase in the high field side indicate negative anisotropy in the ML [17].

The line shape of FMR signal is sensitive to magnetic interactions in the material and hence a change in line shape is an indication of change in the magnetic interactions in the ML. Many a times the observed single FMR signal may arise from overlapping of 2 or more FMR signals.

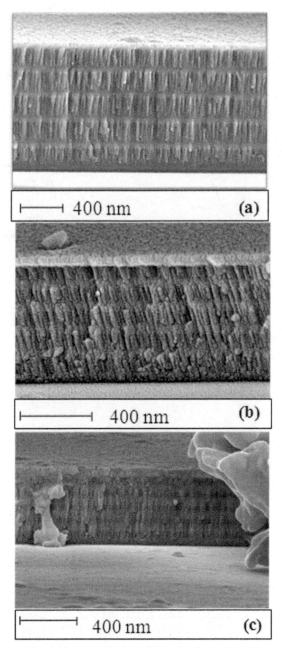

Figure 1. FESEM images of [SnO$_2$ (46nm) CZF (x nm)]$_5$ for x = 249 (a), 83(b) and 42 nm (c).

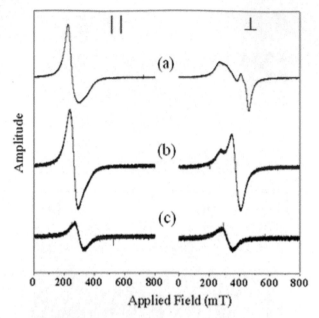

Figure 2. RT FMR spectra of [SnO$_2$ (46 nm) CZF(x nm)]$_5$ in the parallel and perpendicular configuration for x = 249 (a), 83(b) and 42 nm (c).

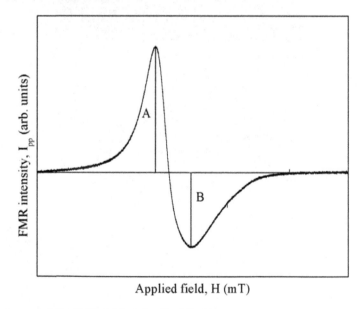

Figure 3. A representative FMR plot depicting the A/B ratio.

The asymmetry parameter, A/B ratio, was estimated by dividing the intensity of the upper part of the FMR signal by that of the lower part (figure 3). A/B > 1 indicates that the intensity of the upper half of the FMR signal is higher than the lower one and vice versa. A/B = 1 indicates a signal with both its upper and lower half with equal intensity.

The A/B ratio at room temperature in the parallel configuration is 1.27, 1.12 and 0.69 for x = 249, 83 and 42 nm respectively. A/B > 1 gives Dysonian line shape to the spectrum which results from lower skin depth and higher conductivity of the sample [18]. Hence the possibility of occurrence of dispersive component due to diffusion of electrons into and out of the skin region cannot be ruled out for the ML with x = 249 and 83 nm.

The asymmetric nature of the FMR spectra of x = 249 nm may also be attributed to the interlayer coupling in the ML. The asymmetry decreases with decrease in CFZ layer thickness presumably due to the decrease in the interlayer coupling strength between the layers. As a consequence of interlayer coupling, the coupled spins precess out of phase with the uncoupled ones. Hence the FMR spectra of spins which precess at different frequencies overlap to give an asymmetric signal [19]. The peak to peak FMR signal intensity (I_{PP}) is measured between the positive and negative peak points along the y-axis and is proportional to the power absorbed. In the parallel configuration I_{PP} values, normalized to the volume of the CZF layers, increases as the magnetic layer (CZF) thickness increases from x = 42 to 83 nm followed by a decrease for x = 249 nm. The resonance field (H_{res}) decreases with increase in the CFZ layer thickness in this geometry. This is attributed to the increase in internal field with increasing CZF layer (magnetic layer) thickness because of which the magnetization also increases. The FMR parameters for both parallel and perpendicular configuration are listed in table 1.

The intrinsic line shape and linewidth of an FMR signal is sensitive to variation of magnetization at the interfaces, surface pits, grain boundaries etc. The peak to peak linewidth (ΔH) is sensitive to inhomogeneity, surface roughness, internal field etc. ΔH values for the ML reported in this study initially decreases from 631 to 541 Oe as x changes from 42 to 83 nm followed by an increase to 704 Oe for x = 249 nm. The ratio of the active layer thickness to dead layer thickness is low for ML with x = 42 nm. As x changes from 42 to 83 nm, the contribution from the dead layer becomes relatively negligible due to large active layer thickness. The increase in ΔH for x = 249 nm may be due to the presence of interlayer coupling [20].

The ML with x = 249 nm exhibits multiple resonances in $\theta_H = 0°$ configuration at RT (figure 2). The resonances occur at 363 and 418 mT for this ML. The occurrence of multiple resonances may be attributed to the excitation of spin waves due to the inter layer coupling of CZF layers mediated by the non-magnetic SnO₂ layers [21]. This implies that the entire ML is coupled by the interlayer exchange coupling as a single magnetic entity. The spin waves propagate through the SnO₂ layers and are sustained by the entire multilayer film. This confirms that for x = 249 nm the CZF layer thickness is ample enough to polarize SnO₂ layer. With decreasing x it loses its capacity to polarize SnO₂ layer. The occurrence of multiple resonances may also be attributed to the existence of two different magnetic phases pertaining to the interfacial region and the bulk of the magnetic layer.

The number of resonances decreases with decreasing CZF thickness in the perpendicular geometry. As the thickness of CZF layer increases, the ML provides sufficient length for the spin wave modes to be formed.

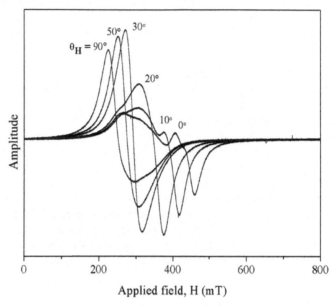

Figure 4. Room temperature FMR spectra of [SnO$_2$(46 nm)CFZ(249 nm)]$_5$ at different orientations.

The FMR spectra of [SnO$_2$(46 nm)/CZF(249 nm)]$_5$ recorded at different orientations is shown in figure 4. For $\theta_H > 20°$ there is only single resonance in the FMR spectra. Small shoulder appears at $\theta_H = 20°$. The multiple resonances appear at $\theta_H < 20°$. The relative intensity of the FMR signals is also angle dependent. As θ_H approaches 0°, the amplitude of the shoulder peaks increases at the cost of that of the main peak. The separation between the resonances also increases and the intensity of the FMR signal decreases as θ_H approaches 0°. This may be due to the prominent demagnetization effects in the perpendicular configuration. A similar behavior is exhibited by the ML with x = 86 nm. There are no multiple resonances for the ML with x = 42 nm.

Figure 5 is a representative plot of the angular dependence of A/B ratio. A/B is close to 2.1 for the perpendicular configuration and decreases as the ML approaches parallel configuration. A/B = 1 for $\theta_H \sim 30°$ showing that the spectra is symmetric at an orientation of 30° of the sample normal with the applied field. Below 30° A/B is < 1 and above 30° A/B is > 1 evidencing the variation in the intensity of the lower and upper part of the FMR spectra with θ_H.

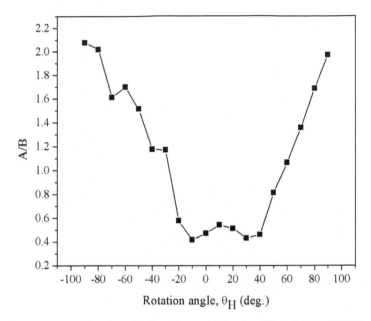

Figure 5. A representative plot of angular dependence of A/B ratio of [SnO₂(46 nm)CFZ(249 nm)]₅ ML.

Figure 6 depicts the angular variation of H_{res}, ΔH and I_{pp}. For all the ML there is no significant change in H_{res} and ΔH the FMR spectra for $\theta_H \geq 60°$. Below 60°, H_{res} and ΔH increase continuously. This shows that the ML is anisotropic in nature. The I_{pp} value initially increases from 0° till 40° followed by a decrease with further increase in θ_H. Thus the ML is homogeneous when placed at an orientation of 40°.

When an applied field is oriented along the film plane, H_{res} increases with decreasing thickness of the CZF layers. But H_{res} decreases with decreasing thickness of the CZF layers when the external magnetic field is oriented perpendicular to the film plane. This behavior is attributed to the decrease of the perpendicular anisotropy energy with decreasing CZF thickness [22]. The difference between the resonance fields in the parallel and perpendicular configuration decreases with decreasing CZF thickness evidencing the fact that the ML with x = 42 nm is relatively less anisotropic compared to the ML with x = 249 and 83 nm.

In the perpendicular configuration, the ML with x = 249 nm exhibits multiple resonances presumably due to the presence of spin waves. The spin wave resonance spectrum at RT and HT for ML with x = 249 nm were initially analyzed using the dispersion relation developed by Kittel as follows [23]

$$H_n = H_0 - \Lambda M_0 \left(\frac{\pi}{L}\right)^2 n^2 \tag{1}$$

where $\Lambda = \frac{2A}{M_0{}^2}$

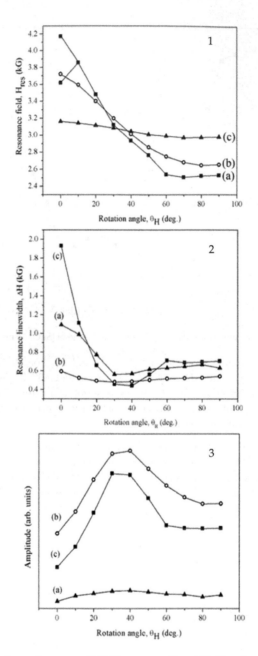

Figure 6. Angle dependence of resonance field (1), resonance linewidth (2) and amplitude (3) of $[SnO_2(46\ nm)CFZ(x\ nm)]_5$ for $x = 249$ (a), 83(b) and 42 nm (c).

Here H$_n$ is the position of the nth mode of the spin wave resonance, H$_0$ is the position of the FMR resonance, L is the thickness of the sample, M$_0$ is the magnetization of the film, n is an odd integer and A is the exchange constant. This relation is valid for uniformly magnetized film. The results obtained do not fit well into the Kittel model where H$_n$ is plotted against n^2. They rather exhibit a linear behavior with n.

According to Portis model [24] for films with non uniform magnetic profile, the magnetization shows a parabolic drop away from the center of the film. The spin wave modes are given by the following relation

$$H_n = H_0 - \frac{4M_0}{L}(4\pi\epsilon)^{1/2}\left(n + \frac{1}{2}\right)$$ (2)

Here ϵ is the distortion parameter in the film.

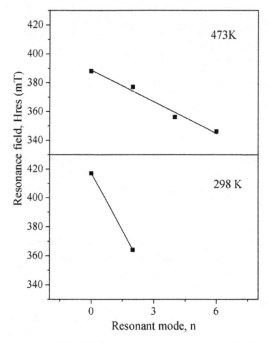

Figure 7. Variation of resonance field with the mode number according to Portis model.

Portis considered that the pinning in the film arises due to non uniform magnetization in the film. This model proposes a linear relation between H$_n$ and n. Also from the above equation, the difference between the resonance modes decreases with decrease in magnetization.

Figure 7 shows the linear fits corresponding to Portis model. The exchange constant is proportional to the slope of the line. The slope of the line decreases with increasing temperature. Hence the exchange constant decreases with increasing temperature leading to weak coupling and hence lower magnetization. Also, the separation between the resonance modes decreases at higher temperatures, indicating a decreasing magnetization.

The effective magnetization ($4\pi M_{eff}$) and effective g value (g_{eff}) calculated using the Kittel's relations are listed in table1. At room temperature $4\pi M_{eff}$ increases and g_{eff} decreases with increasing CZF thickness. This may be due to increasing FM layer thickness [25]. The various factors which can account for deviation of g_{eff} from that of the free electron are spin orbit coupling, coupling at the interfaces etc [26].

x (nm)	H_{res} (kG)		ΔH_{res} (kG)	g (± 0.01)		$4\pi M_{eff}$ (Oe)	g_{eff}	ΔH (G)	
	$\theta_H = 90°$	$\theta_H = 0°$		$\theta_H = 90°$	$\theta_H = 0°$			$\theta_H = 90°$	$\theta_H = 0°$
42	2.977	3.165	0.188	2.21	2.08	125.76	2.16	631	750
86	2.649	3.720	1.071	2.48	1.77	728.39	2.19	541	590
249	2.526	3.633, 4.173	1.107, 1.647	2.60	1.81, 1.58	754.41, 1132.88	2.28, 2.14	703	1930

Table 1. FMR parameters for [SnO2(46 nm)/ CZF (x)]5 ML for various values of x.

The perpendicular anisotropy field H_K is obtained from the following equation [27]

$$H_K = 4\pi M_S - 4\pi M_{eff} \tag{3}$$

The values of K are listed in table 2. In multilayers, there are two factors which contribute to the effective anisotropy – anisotropy due to the interfacial region (K_S) and anisotropy due to the bulk volume of the magnetic layer (K_V). The relation is given as follows [28]

$$K = K_V + 2\frac{K_S}{x} \tag{4}$$

Where x is the thickness of the magnetic layer. The factor of 2 in the second term on the (r. h.s) of the equation is due to an assumption that the magnetic layer is bound by identical interfaces on either side. When (K.x) is plotted against t (figure 8), the slope gives the value of Kv and the intercept gives the value of Ks. The negative slope indicates a negative volume anisotropy which tends to confine the magnetic moment in the plane of the film [29]. The intercept at x = 0 gives positive value of Ks which tends to align the magnetic moment perpendicular to the surface of the film. The intercept on the x axis gives the thickness at which the contribution from Kv outweighs that from Ks. From the figure, the intercept on x axis is found to be 56 nm. Thus below 56nm, major contribution comes from interfaces. The Ks and Kv values obtained from the plot are 194980 and -3671 erg/cm³

The value of K changes from positive to negative as the thickness of CZF increases (figure 8). Negative values of K indicates in-plane anisotropy [30]. Thus with increasing CZF thickness, the direction of magnetization is confined to the plane of the sample surface. The thickness value at which the anisotropy changes sign is 56 nm. The perpendicular anisotropy for small values of x may be due to the lattice mismatch at the interfaces [31].

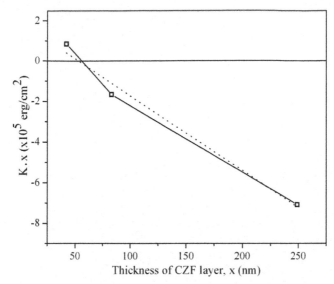

Figure 8. (K.x) vs x plots for [SnO₂(46 nm)CFZ(x nm)]₅ ML for x = 43, 83 and 249 nm.

Gilbert damping is a spin relaxation phenomenon in magnetic systems which controls the rate at which the spins reach equilibrium. Spin-orbit coupling [32], non-local spin relaxations like spin wave dissipation [33,34] and disorder present in the materials are the major factors causing Gilbert damping. The damping parameter α values are 0.0096, 0.0083 and 0.0108 for x = 42, 83 and 249 nm respectively. The increase in α value with CZF thickness indicates the increase in the damping of spins with increasing thickness.

x	α		K_1
(nm)	$\theta_H = 90°$	$\theta_H = 0°$	(10^3 erg/cm³)
42	0.049	0.057	2.027
86	0.041	0.045	-1.989
249	0.053	0.146	-2.845, -1.254

Table 2. Damping and anisotropy parameters for or [SnO₂(46 nm)/ CZF (x)]₅ ML for various values of x.

The temperature dependent FMR studies in the temperature range 133- 473K were carried out on the ML. Figure 9 shows the FMR spectra of the ML at various temperatures. The change in line shape indicates change in the magnetic interaction with increasing temperature.

Figure 10 shows the variation of I_{pp} with temperature in the parallel configuration. I_{pp} increases with increasing temperature for all the ML. I_{pp} of the ML with x= 42 nm exhibits a linear trend. Whereas for ML with x=83 and 249 nm, the increase is not linear. It increases slowly from 133K to 298K (RT), followed by a rapid increase with temperature in the temperature range RT- 473K. From 133K to 250K there is no significant difference in amplitude for ML with x= 83 and 249 nm. At temperatures above RT the amplitude of ML with x= 83 nm is higher than that of ML with x = 249 nm.

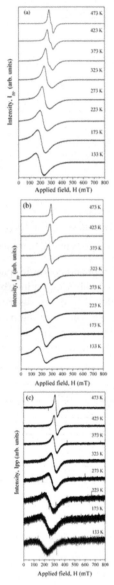

Figure 9. FMR spectra of [SnO₂(46 nm)CFZ(x nm)]₅ ML at various temperatures for x = 249 (a), 83 (b) and 42 nm (c)in the parallel configuration.

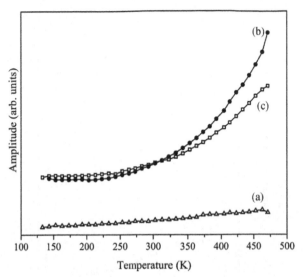

Figure 10. Temperature dependent amplitude of [SnO₂(46 nm)CFZ(x nm)]₅ for x = 249(a), 83(b) and 42 nm (c).

Variation of ΔH with temperature in the parallel configuration is displayed in figure 11. The width of the FMR spectra increases continuously with decreasing temperature for x= 42 nm. For x = 83 nm it decreases with temperature and saturates at ~ 215 K . Whereas for x = 249 nm it begins saturating at ~ RT presumably due to freezing of spins. The H_res increases with increasing temperature (figure 12) in all the ML presumably due to increasing internal field at low temperatures. The ML with x = 249 nm exhibits a linear dependence in the entire temperature range whereas the ML with x = 83 and 42 nm exhibit saturating tendency at higher temperatures in the parallel configuration. This may be an interfacial effect. For x = 42 nm the ratio of spins at the interfaces to the spins in the bulk is high. Whereas for x = 83 and 249 nm the ratio is lower. Hence the contribution from the bulk of the film becomes significant than that from the interfaces. The saturating tendency at high temperatures for the ML with x = 83 and 42 nm shows that even at higher temperatures, the thermal energy is not sufficient to unlock the spins at the interfaces.

Similar behavior is observed for CZF thin films [12]. The increase in saturation temperature with increase in CZF thickness could be due to the interlayer coupling.

The change of slope of ΔH indicates a change in the kind of magnetic interactions in the ML. The line shape of the FMR spectra also changes around the same temperature. The curvature of the temperature dependent of ΔH plot changes as x increases from 42 to 249 presumably due to increase in active to dead layer ratio. Decreasing trend in I_pp and H_res and increasing trend in ΔH with decreasing temperature is a signature of superparamagnetism [35]. The resonance field of a given particle includes contribution from magneto crystalline anisotropy field and demagnetizing field.

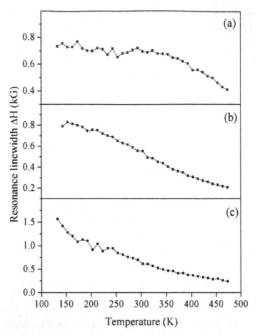

Figure 11. Temperature dependent resonance linewidth of [SnO$_2$(46 nm)CFZ(x nm)]$_5$ for x = 249(a), 83(b) and 42 nm (c).

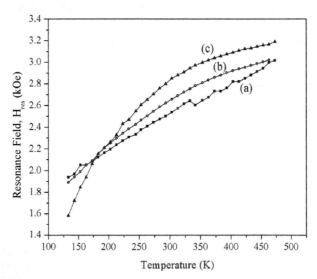

Figure 12. Temperature dependent resonance field, H$_{res}$, in the parallel and perpendicular configuration of [SnO$_2$(46 nm)CFZ(x nm)]$_5$ for x = 249(a), 83(b) and 42 nm (c).

For a system containing randomly oriented particles, the magnetic resonance signal is broad. At high temperatures where k_bT is much greater than the anisotropy barriers the thermal fluctuations of the magnetic moments reduces the angular anisotropy of the resonance fields and the linewidth of individual nanoparticle [36] resulting in a superparamagnetic resonance (SPR) spectra in the case of nanoparticles. At low temperatures the particle moments are unable to overcome the local anisotropy barriers and, thus, become trapped in metastable states (blocking phenomenon). This results in broad resonance spectra with linewidth comparable to that of the resonance field.

A characteristic feature of SPR is that at low temperatures, ΔH is almost equal to the value of H_{res}. For the ML with x = 42 nm around 40% of CZF layer is dead. The diffused region consists of fine particles of CZF dispersed in SnO₂ matrix. Hence the SPR phenomenon is pronounced for this ML. Whereas for ML with x = 83 and 249 nm, the thickness of the diffused region is small compared to that of the bulk CZF layer. Hence the value of ΔH at low temperatures is not same as that of H_{res}.

FMR susceptibility is estimated by calculating the double integrated intensity i.e. area under the absorption curve in FMR is proportional to the concentration of spins. Figure 13 shows temperature dependence of ESR susceptibility in the parallel configuration for [SnO₂(46 nm)CFZ(x nm)]₅ ML. It initially increases with decreasing temperature, reaches a maximum and then decreases with further decrease in temperature evidencing the superparamagnetic behavior. The width of the transition is large due to the wide spread of blocking temperatures.

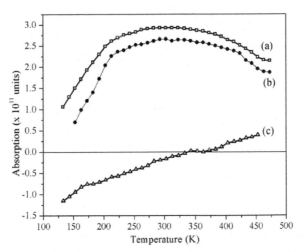

Figure 13. Variation of FMR susceptibility, χ_{FMR} with temperature of [SnO₂(46 nm)CFZ(x nm)]₅ ML for x = 249 (a), 83 (b) and 42 nm (c).

For ML with x = 42 nm, χ_{FMR} decreases continuously with decrease in temperature down till 183 K. The negative absorption is due to the fact that the area under the lower part of the

FMR curve is greater than that of the upper part. The Curie- Weiss law does not fit 1/ χ FMR vs T plot due to the continuous curvature in the entire temperatures range.

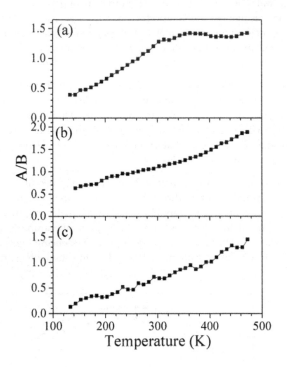

Figure 14. Variation A/B ratio with temperature of [SnO₂(46 nm)CFZ(x nm)]₅ ML for x = 249(a), 83(b) and 42 nm (c).

Figure 14 shows the A/B ratio with temperature for these ML. It decreases almost linearly with decreasing temperature from 400 to 200 K for the ML with x = 42 nm. Below 200 K there is a change in slope. The temperature dependence of A/B ratio exhibits different slopes in different temperature region for the ML with x = 83 nm. Whereas for the ML with x = 249 nm, there is no significant change in the A/B ratio in the temperature range between 500 and 350 K. Below 350 K the ratio decreases continuously. The temperature at which the slope for A/B ratio changes matches with that of temperature dependence of linewidth.

Temperature dependence of FMR spectra of these ML in the perpendicular configurations is displayed in figure 15. The resonance field shifts towards higher values with decreasing temperature. This is accompanied with decrease in intensity and increase in line width. This indicates that AFM interactions dominate as temperature decreases in the perpendicular configuration.

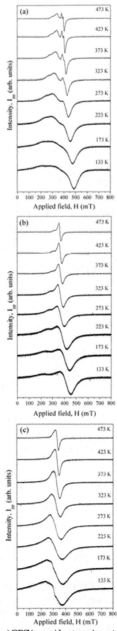

Figure 15. FMR spectra of [SnO₂(46 nm)CFZ(x nm)]₅ at various temperatures in the perpendicular configuration for x = 249 (a), 83 (b) and 42 nm (c).

Figure 15(a) also shows the variation in the positions of the multiple resonances for x = 249. They are narrow and closely spaced at 473 K. As temperature decreases, the FMR peaks broaden and move away from each other with decreasing temperature. The peak at the lower field moves further towards lower field and the one at the higher field moves further towards higher field. This is a clear indication of existence of both FM and AFM phase. The relative intensity of the peak at higher field decreases and that of the one at lower field increases with decreasing temperature. The distance between the peaks increases with decreasing temperature. This indicates increase in phase separation with decrease in temperature.

I_{pp} value in the perpendicular configuration is lower than that in the parallel configuration at all temperatures. However, the trend remains the same in both parallel and perpendicular configurations. The ΔH value in the perpendicular configuration is not very much different from that in the parallel configuration for the ML with x = 42 and 83 nm (figure 16). On the other hand, ΔH for x = 249 ML is much higher in the perpendicular configuration.

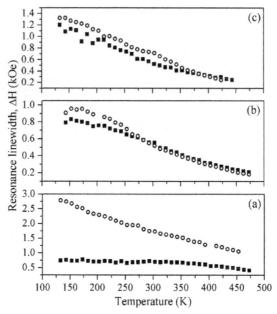

Figure 16. Temperature dependent resonance linewidth in the parallel (full circle) and perpendicular (open circle) configuration of [SnO₂ (46 nm)CFZ(x nm)]₅ for x = 249(a), 83(b) and 42 nm (c).

Figure 17 shows the variation of H_{res} in the parallel and perpendicular configuration. Unlike the parallel configuration, the resonance field increases with decreasing temperature in the perpendicular configuration for ML with x = 249 and 83 nm. H_{res} remains insensitive to temperature in the range 300- 473K. Below 300 K it increases with further decrease in temperature.

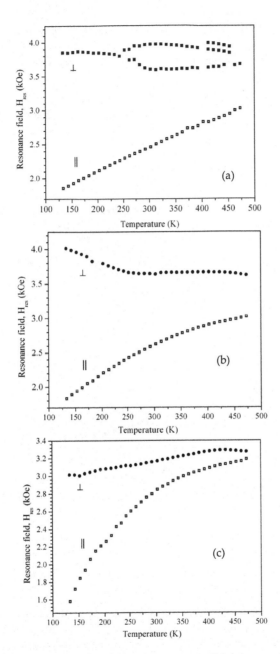

Figure 17. Temperature dependent resonance field, H_{res}, in the parallel and perpendicular configuration of $[SnO_2(46\ nm)CFZ(x\ nm)]_5$ for x = 249 (a), 83 (b) and 42 nm (c).

Whereas for the ML with x = 42 nm, initially there is a small increase in H_{res} as temperature decreases from 473 to 433 K. Below 433 K it decreases continuously. At all temperatures, H_{res} is less sensitive to temperature in the perpendicular configuration than in the parallel configuration.

For all the ML, the FMR susceptibility increases with decreasing temperature in the perpendicular configuration. The effective magnetization is a measure of net magnetization within the ML. Figure 18 shows the variation of effective magnetization with temperature. It increases with decreasing temperature for all ML.

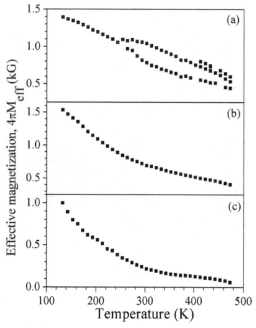

Figure 18. Variation of effective magnetization with temperature for [SnO$_2$ (46 nm)CFZ(x nm)]$_5$ ML for x = 249(a), 83(b) and 42 nm (c).

Figure 19 is a representative graph of [SnO$_2$ (46 nm)/ CZF(249 nm)]$_5$ ML showing the variation of FMR susceptibility with temperature in both parallel and perpendicular configurations. Unlike the parallel configuration, χ_{FMR} in the perpendicular configuration increases continuously with decreasing temperature.

The FMR spectra were recorded at different orientations of the normal to the sample with the applied field (θ_H). The rotational dependence of FMR spectra was measured at 133, 298 (room temperature) and 473 K. At all temperatures the resonance field increases continuously as θ_H decreases exhibiting uniaxial anisotropy. For the ML, the difference between ΔH values in the parallel and perpendicular configuration increases with decreasing temperature. This shows increasing anisotropy with decreasing temperature. The multiple resonance for the ML with x = 249 is prominent at 298 and 473 K (figure 20(a)). For ML with x < 249 nm, the absence of multiple resonances even at high temperatures indicates absence of interlayer coupling. At all temperatures the ΔH_{res} is highest for the ML with x = 249 nm. These multiple resonances disappear with decreasing temperature.

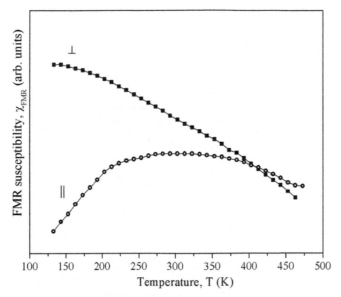

Figure 19. Representative plot of the FMR susceptibility, FMR of [SnO₂ (46 nm)CFZ(249 nm)]₅ ML in parallel and perpendicular configuration.

At room temperature, ΔH shows a minimum ~ 30-40° (figure 21). I_{pp} shows maximum around the same value. Hence the sample has an easy axis of magnetization at that inclination. At 473 and 298 K its intensity initially increases, reaches a maximum and then decreases with decreasing θ_H. Whereas at 133 K, I_{pp} values remains constant initially then decreases continuously with decreasing value of θ_H.

Another evidence for decreasing anisotropy with decreasing x is shown in figure 21. At 473 K, ΔH for the ML with x = 249 nm initially increases, exhibits a minimum ~ 40° and then increases continuously with further increase in θ_H. Whereas for the ML with x = 83 nm, with increasing θ_H, ΔH value remains constant till certain angle then decreases continuously. On the other hand the ΔH value for the ML with x = 42 nm, the ΔH value continuously decrease reaches a minimum and then increases with increasing θ_H. The occurrence of minimum in ΔH may be due to existence of regions with different effective magnetization values due to inhomogeneous nature of the sample [37]. The minimum in ΔH shifts towards higher angle with decreasing x. Similar behavior is observed at 298 K for all the ML. With decreasing

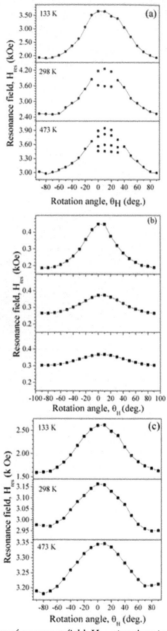

Figure 20. Rotational dependence of resonance field, H_{res}, at various temperatures for [SnO$_2$ (46 nm)CFZ(x nm)]$_5$ ML where x = 249(a), 83(b) and 42 nm (c).

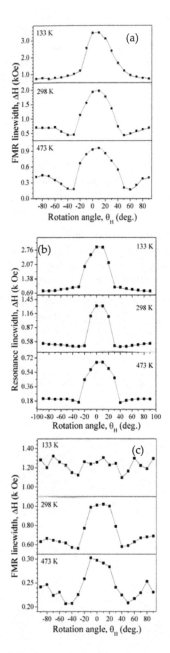

Figure 21. Rotational dependence of resonance linewidth, ΔH, at various temperatures for [SnO₂ (46 nm)CFZ(x nm)]₅ ML for x = 42(a), 83(b) and 249 nm (c).

temperature, the curvature of the angle dependent linewidth changes. At 133 K, for x = 249 nm, the minima vanishes and ΔH increases continuously with increasing θ_H. For the ML with x = 83 nm, the curvature is reversed at 133 K. It increases with increasing θ_H, attains a maximum value at ~ 40° and then decreases. On the contrary there is no significant change in ΔH for the ML with x = 42 nm. Thus at 133 K, x = 249 nm ML is more anisotropic than the ML with x = 42 and 83 nm.

3.2. Effect of spacer layer thickness

The effect of thickness of spacer (SnO2) layer was studied on 3 sets of samples with varying thickness of the SnO2 layer. The total number of bilayers is 5 and individual CZF layer thickness is 83 nm for all the samples. The ML samples were [SnO2(x nm) CZF(83 nm)]5 ML for x = 46 , 115 and 184 nm .

Stack of alternate dark and light layers are evident from the FESEM images (figure 22). The dark columnar layers correspond to CZF and the light layers correspond to SnO2. The variation in the thickness ratio between SnO2 and CZF is also clearly visible from these images. The absence of fringes in XRR spectra may be due to large thickness of the ML stack.

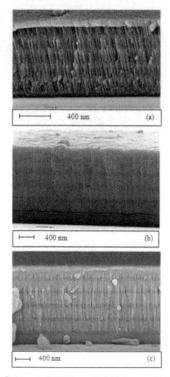

Figure 22. FESEM images of [SnO2(x nm) CZF(83 nm)]5 ML for x = 46 (a), 115 (b) and 184 nm (c).

FMR studies were carried out at different temperatures and by varying the angle θ_H between the film normal and the direction of the applied field. Figure 23 shows the first derivative of the FMR absorption spectra of the ML at room temperature in parallel (θ_H = 90°) and perpendicular (θ_H = 0°) geometry, normalized to the volume of the CZF layers. I$_{pp}$ increases as the spacer (SnO$_2$) thickness increases from x = 46 to 115 nm followed by a decrease for x = 184 nm. The peak to peak linewidth (ΔH) is sensitive to inhomogenieties, surface roughness, internal field etc. The χ FMR is estimated by calculating double integrated intensity (DI). Both ΔH and DI follow a trend similar to that of I$_{pp}$ with increase in spacer layer thickness. This behavior may be attributed to the oscillation in the exchange coupling across the NM layer due to quantum interference of the electron waves reflected from the interfaces. The A/B ratio is 0.82, 0.74 and 0.79 for x = 46, 115 and 184 nm respectively. The asymmetry ratio is not very different from each other. For the ML with x = 115 nm, the upper part of the FMR signal rises steeply whereas the rise is relatively slow for the lower part indicating a negative anisotropy.

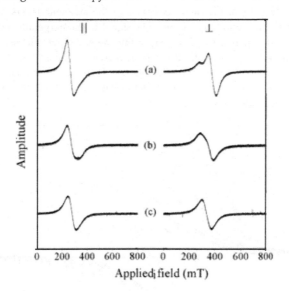

Figure 23. Room temperature FMR spectra of [SnO$_2$(x nm) CZF(83 nm)]$_5$ ML in the parallel and perpendicular configuration for x = 46 (a), 115 (b) and 184 nm (c)..

The resonance field (H$_{res}$) increases from 2.65 to 2.78 kG and the g value decreases from 2.47 to 2.36 with increase in SnO$_2$ layer thickness in parallel geometry. The shift in H$_{res}$ to higher values might be due to the decrease in interlayer coupling strength. The H$_{res}$ decreases and g values increases from 1.77 to 1.96 with increasing spacer layer thickness in the perpendicular geometry. This might be attributed to lower perpendicular anisotropy for well separated films [38]. The FMR parameters in the parallel and perpendicular configurations are listed in table 3. The effective magnetization decreases with increasing SnO$_2$ thickness.

x (nm)	H_{res} (kG)		ΔH_{res} (kG)	g (±0.01)		$4\pi M_{eff}$ (Oe)	ΔH (G)		M_s (emu/cc CZF)
	$\theta_H = 90°$	$\theta_H = 0°$		$\theta_H = 90°$	$\theta_H = 0°$		$\theta_H = 90°$	$\theta_H = 0°$	
46	2.652	3.721	1.069	2.48	1.77	710	541	595	50
115	2.733	3.452	0.719	2.41	1.90	485	693	733	100
184	2.784	3.356	0.576	2.36	1.96	406	675	736	30

Table 3. FMR parameters of [SnO$_2$(x nm) CZF(83 nm)]$_5$ multilayers for various spacer layer thickness

The ML with x = 46 nm exhibits multiple peaks in the FMR spectrum recorded in the perpendicular configuration. This may be due to the interlayer coupling of the CZF layers mediated by the non magnetic spacer layer. When two spins are coupled across the spacer layer, they tend to precess at a frequency different from that of the rest. Hence the resonance for the coupled spins occurs at a different field. The absence of multiple splitting in the other two samples indicates lack of interlayer coupling in the magnetic layers. Apart from the multiple resonances, the asymmetry in the FMR signal in the perpendicular configuration also decreases with the increase in SnO$_2$ layer thickness. As the thickness of the spacer increases, the distance between the CZF layers increases and hence the coupling between them decreases leading to symmetrical FMR spectra. The effective magnetization value at room temperature was calculated using the Kittel's relations and are 710, 485 and 406 Oe for x = 46, 115 and 184 nm respectively.

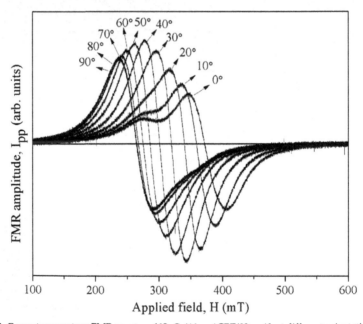

Figure 24. Room temperature FMR spectra of [SnO$_2$(46 nm)CFZ(83 nm)]$_5$ at different orientations.

The angle dependent FMR spectra are shown in figure 24. The symmetry in the FMR spectra increases as θ_H decreases from 90° to 40°. For $\theta_H > 40°$ the spectra becomes asymmetric and the single FMR peak splits into two resonance peaks. With decreasing θ_H the separation of the peaks increases.

Figure 25 is a representative graph of the angular dependence of ΔH and I_{pp}. For all ML, ΔH value initially decreases till $\theta_H = 40°$ and then increases till $\theta_H = 0°$. Whereas, I_{pp} increases till $\theta_H = 40°$ and then decreases till $\theta_H = 0°$. This indicates that at $\theta_H = 40°$, the susceptibility of the film increases resulting in higher magnetization. This may be due to existence of regions with two different effective magnetizations. For all the samples H_{res} shifts to higher fields as the angle between the sample normal and the direction of the field (θ_H) changes from 90° to 0°. This indicates that the ML exhibits uniaxial anisotropy.

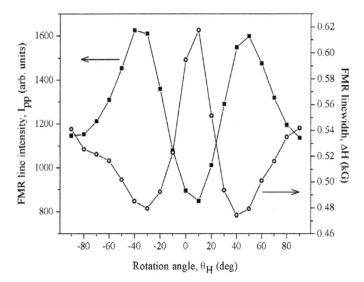

Figure 25. Variation of room temperature FMR linewidth (ΔH) and intensity (I_{pp}) of [SnO₂ (46 nm) CZF(83 nm)]₅ ML at different orientations of the film.

FMR spectra recorded at various temperatures between 133 and 473 K are shown in figure 26. The change in the lineshape with decreasing temperature evidences an evolution of a different type of magnetic interaction.

The FMR signal is asymmetric at high temperatures and the symmetry increases with decreasing temperature. In the high temperature region A/B is greater than 1 indicating that the upper part of the FMR spectrum is of higher intensity than the lower half and indicates the presence of dispersive component in the ML. Figure 27 shows the variation of A/B with temperature.

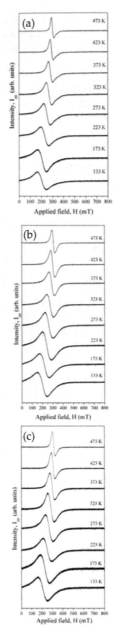

Figure 26. FMR spectra of [SnO₂ (x nm) CZF(83 nm)]₅ ML recorded at various temperatures in parallel configuration for x = 46 (a), 115 (b) and 184 (c).

The I_{PP} value decreases with decreasing temperature. The increase in ΔH at lower temperatures indicates the large spin relaxation times and hence large coercivity at low temperatures (figure 28(a)).

The decreasing trend in I_{PP} and H_{res} and increasing trend in ΔH with decreasing temperature is a signature of superparamagnetism. Figure 29 shows variation of FMR susceptibility, χ FMR as a function of temperature. It initially increases, reaches a maximum and then decreases with decreasing temperature confirming the superparamagnetic nature of the ML. Figure 30 shows the temperature dependent FMR spectra in the perpendicular configuration. All the ML samples exhibit multiple resonances. The resonances broaden and the separation between them increases with decreasing temperature. The intensity of the peak at lower field side increases and that at higher field side decreases with decreasing temperature.

The behavior of ΔH and I_{PP} is same as that of $\theta_H = 90°$. The difference between ΔH in both the configurations is negligible in the HT region for all the samples. The substantial difference occurs at low temperatures (figure 31(a)). This may be due to the presence of magnetic inhomogeneities in the ML [39]. In the parallel configuration H_{res} decreases with decrease in temperature indicating strengthening of FM interactions. Whereas in the perpendicular configuration it is less sensitive to temperature in the HT- RT range, then increases slowly with decrease in temperature (figure 31 (b)).

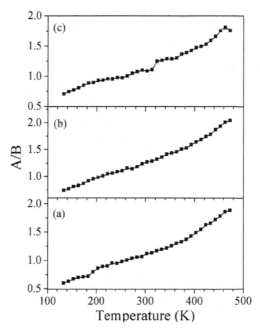

Figure 27. Variation of A/B ratio of [SnO2 (x nm) CZF (83 nm)]5 with temperature for x = 46 (a), 115 (b) and 184 (c).

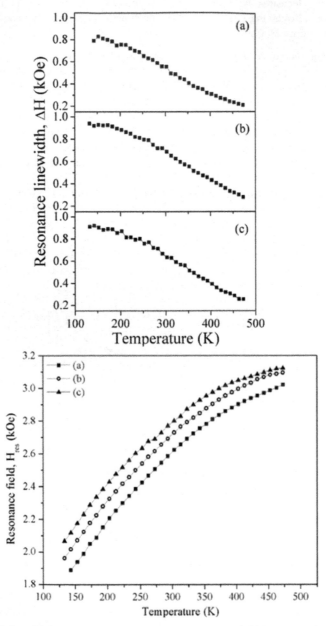

Figure 28. Variation of FMR resonance linewidth (top) and resonance field (bottom) with temperature of [SnO₂ (x nm) CZF(83 nm)]₅ for x = 46 (a), 115 (b) and 184 nm (c).

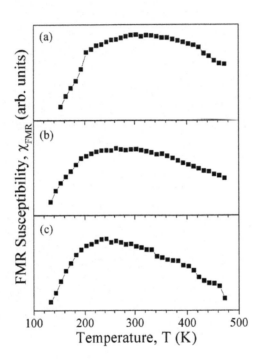

Figure 29. Variation of FMR susceptibility, χ_{FMR} with temperature of [SnO$_2$(x nm)CFZ(83 nm)]$_5$ ML for x = 46 (a), 115 (b) and 184 nm (c).

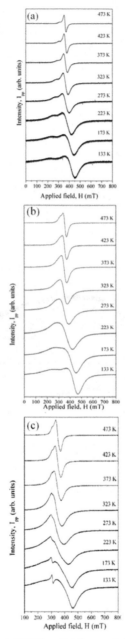

Figure 30. FMR spectra of [SnO₂ (x nm) CZF(83 nm)]₅ ML recorded at various temperatures in perpendicular configuration for x = 46 (a), 115 (b) and 184 (c).

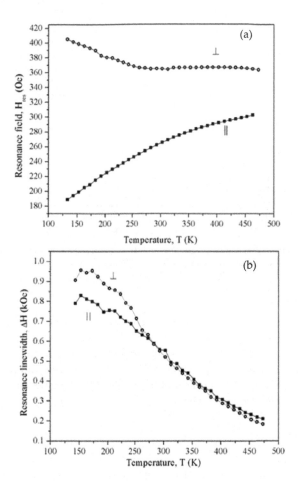

Figure 31. Variation of resonance field, H$_{res}$ (a) and resonance linewidth, ΔH (b) with temperature in parallel and perpendicular configurations for [SnO₂ (46 nm) CZF(83 nm)]₅ ML .

The effective magnetization is found to increase with decreasing temperature. Major contribution to the effective magnetization comes from the perpendicular anisotropy. This implies that as the temperature decreases the perpendicular anisotropy relaxes. χ$_{FMR}$ shows a different behavior as a function of temperature in the perpendicular configuration. Figure 32 is a representative plot of variation of χ$_{FMR}$ with temperature in parallel and perpendicular configuration. It increases continuously with decreasing temperature for all the ML in the perpendicular configuration.

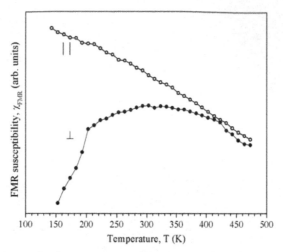

Figure 32. Representative plot showing variation of FMR susceptibility with temperature in parallel and perpendicular configurations for [SnO₂ (46 nm) CZF(83 nm)]₅ ML .

4. Conclusions

The temperature dependent ferromagnetic resonance (FMR) studies on [SnO₂/ Cu-Zn ferrite]ₙ as function of magnetic layer and spacer layer thickness were carried out in the temperature range 100- 475 K. The study of temperature dependence of FMR lineshape, peak-to-peak linewidth, peak-to –peak line intensity and resonance field provide an insight into the interfacial effects. The inplane and out of plane FMR studies provide information about the magnetic anisotropies of the ML.

Author details

R.Singh and S.Saipriya
School of Physics, University of Hyderabad, Central University P.O., Hyderabad, India

5. References

[1]   P. Bruno and C. Chappert, *Phys. Rev. Lett. B* 67 (1991) 1602

[2]   D.M.Edwards, J.Mathon, R.B.Muniz, and M.S. Phan, *Phys. Rev. Lett.* 67 (1991) 493

[3]   C.H.Marrows, Nathan Wiser, B. J. Hickey, T.P.A Hase and B.K. Tanner, *J. Phys.: Cond. Matter* 11 (1999) 81

[4]   K. R. Nikolaev, A. Yu. Dobin, I. N. Krivorotov, W. K. Cooley, A. Bhattacharya, A. L. Kobrinskii, L. I. Glazman, R. M. Wentzovitch, E. Dan Dahlberg, and A. M. Goldman *Phys. Rev. Lett.* 85 (2000) 3728

[5]   D. M. Lind, S. D. Berry, G. Chern, H. Mathias, and L. R. Testardi, *Phys. Rev. B* 45 (1992) 1838.

[6]   M. A. James, F.C. Voogt, L. Niesen, O.C. Rogojanu and T. Hibma, *Surf. Sci.* 402 (1998) 332

[7]   P. J. Van der Zaag, R.M. Wolf, A.R. Ball, C. Bordel, L.F. Feiner and R. Jungblut, *J. Magn. Magn. Mater.*148 (1995) 346

[8]   S. Saipriya, Joji Kurian and R. Singh, *IEEE Trans Mag.*147 (2011) 10

[9]   S. Saipriya, Joji Kurian and R. Singh, AIP Conf. Proc. 1451 (2012) 67

[10] S. Saipriya and R. Singh, *Mater. Lett.* 71 (2012) 157

[11] S. Saipriya, Joji Kurian and R. Singh, J. Appl. Phys 111 (2012) 07C110

[12] M. Sultan and R. Singh, J. Appl. Phys. 107 (2010) 09A510

[13] S. Saipriya, M. Sultan and R. Singh, Physica B 406 (2011) 812

[14] M. Sultan and R. Singh, *J. Phys. D: Appl. Phys.* 42 (2009) 115306

[15] G. S. Bales and A. Zangwill, J. Vac. Sci. Technol A 9 (1991) 145

[16] O. S. Josyulu and J. Shobanadri, *J. Mat. Sci.* 20 (1985) 2750

[17] Janhavi P. Joshi, Rajeev Gupta, A. K. Sood and S. V. Bhat, *Phys. Rev. B* 65 (2001) 024410

[18] Y. Gong, Z. Cevher, M. Ebrahim, J. Lou, C. Pettiford, N. X. Sun and Y. H. Ren, *J. Appl. Phys.* 106 (2009) 063916

[19] S. S. Yan, *Acta Metullurgica Sinica* 9 (1996) 283

[20] A. Ohtomo, D. A. Muller, J. L. Grazul and H. Y. Hwang, *Nature* 419 (2002) 378

[21] S.S. Kang, J.W. Feng, G.J. Jin, M. Lu, X.N. Xu, A. Hu, S.S. Jiang and H. Xia, *J. Magn. Magn. Mater.* 166 (1997) 277

[22] C. Kittel, *Phys. Rev.* 110 (1958) 1295

[23] A. M. Portis, *Appl. Phys. Lett.* 2 (1963) 69

[24] S. J. Yuan, L. Wang, R. Shan and S.M. Zhou, *Appl. Phys. A* 79 (2004) 701

[25] R. D. McMichael, M. D. Stiles, P. J. Chen, and W. F. Egelhoff, Jr., *Phys. Rev. B* 58 (1998) 8605

[26] M. Sultan and R. Singh, J. Phys: Conf. Ser. 200, 072090 (2010).

[27] U. Gradmann and A Mueller, *Phys. Status. Solidi.* 27 (1968) 313

[28] M. T. Johnson, P. J. H. Bloemen, F. J. A. den Broeder and J. J. de Vriesy, *Rep. Prog. Phys.* 59 (1996) 1409

[29] J. H. Jung, S. H. Lim and S. R. Lee, *J. Appl. Phys.* 108 (2010) 113902

[30] J. Pelzl et al, *J. Phys.: Condens. Matter* 15 (2003) S451

[31] M. C. Hickey and S. J. Moodera, *Phys. Rev. Lett.* 102 (2009) 137601

[32] Y. Tserkovnyak, Arne Brataas, Gerrit E. W. Bauer and Bertrand I. Halperin, *Rev. Mod. Phys.* 77 (2005) 1375

[33] G. Eliers, M. Lüttich and M. Münzenberg, *Phys. Rev. B.* 74 (2006) 054411

[34] R. Rai, K. Verma, S. Sharma, Swapna S. Nair, M. A. Valente, A. L. Kholkin and N. A. Sobolev, *J. Phys. Chem. Solids* 72 (2011) 862

[35] M. Marysko and J. Simsova, *Phys. Stat. Sol. A,* 33 (1976) K133-K136

[36] M. R. Diehl, J-Y Yu, J. R. Heath, G. A. Held, H. Doyle, S. Sun and C. B. Murray, *J. Phys. Chem. B* 105 (2001) 7913

[37] R. Topkaya, M. Erkovan, A. Öztürk, O. Öztürk, B. Aktaş and M. Özdemir, *J. Appl. Phys.*108 023910
[38] B. X. Gu and X. Wang, *Acta Metalurgica Sinica(English letters)*12 (1999) 181

Chapter 4

Instrumentation for Ferromagnetic Resonance Spectrometer

Chi-Kuen Lo

Additional information is available at the end of the chapter

1. Introduction

Even FMR is an antique technique, it is still regarded as a powerful probe for one of the modern sciences, the spintronics. Since materials used for spintronics are either ferromagnetic or spin correlated, and FMR is not only employed to study their magneto static behaviors, for instances, anisotropies [1,2], exchange coupling [3,4,5,6], but also the spin dynamics; such as the damping constant [7,8,9], g factor [8,9], spin relaxation [9], etc. In this chapter a brief description about the key components and techniques of FMR will be given. For those who have already owned a commercial FMR spectrometer could find very helpful and detail information of their system from the instruction and operation manuals. The purpose of this text is for the one who want to understand a little more detail about commercial system, and for researchers who want to build their own spectrometers based on vector network analyzer (VNA) would gain useful information as well.

FMR spectrometer is a tool to record electromagnetic (EM) wave absorbed by sample of interest under the influence of external DC or Quasi DC magnetic field. Simply speaking, the spectrometer should consist of at least an EM wave excitation source, detector, and transmission line which bridges sample and EM source. The precession frequency of ferromagnetics lies at the regime of microwave (μ-wave) ranged from 0.1 to about 100 GHz, therefore, FMR absorption occurs at μ-wave range. The generation, detection and transmission at such this high frequency are not as simple as those for DC or low frequency electronics. According to transmission line and network theories [10,11], impedance of 50 Ω between transmission line and load has to be matched for optimization of energy transfer. FMR spectrometer also has a resonator and an electro magnet which produces magnetic field to vary the sample's magnetization during the measurement. Sample which is mounted inside the cavity absorbs energy from the μ-wave source. The detector electronics records the changes on either the reflectance or transmittance of theμ -wave while magnetic field is being swept [12]. Most commercial spectrometers, for examples, ELEXSYS-II E580, Bruker Co [13], and JES-FA100, JOEL Ltd. [14], character FMR by measuring the reflectance at fixed band frequency. μ-wave is generated by a Klystron, which goes to a metallic cavity via a wave guide. Signal reflected back from the cavity through the same waveguide to a detector, the Schottky barrier diode. Noted that a circulator is employed, such that the μ-wave does go to the detector and not return back to the generator. The integration of the source, detector, circulator, protected electronics, etc. in a single box, is named "Microwave Bridge". The basic configuration of most FMR spectrometer is shown in Fig.1. FMR absorption spectrum is obtained by comparing the incident and reflected signals. However, the signal to noise ratio (SNR) given by this kind of reflectometer is still not large enough to recognize the spectrum, and lock-in technique is needed to enhance SNR to acceptable value, for example, bigger than 5.

The operational frequency of the source, detector and cavity cannot be varied broadly, therefore, it is necessary to change the microwave bridge and cavity while working with different band frequency. Dart a glance at a FMR spectrometer, the key parts are the microwave bridge, cavity, gauss meter, electromagnet, and lock-in amplifier for signal process. Surely, automation and data acquisition are also crucial.

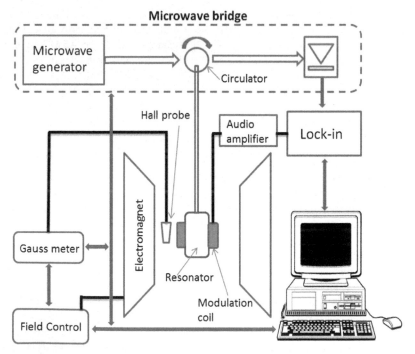

Figure 1. Basic configuration of FMR spectrometer. The microwave bridge mainly consists of the microwave generator, circulator, and detector.

2. Key components of FMR

2.1. Metallic cavity

A Metallic resonator (cavity) is a space enclosed by metallic walls which sustain electromagnetic standing wave or electromagnetic oscillation. The cavity has to be coupled with external circuit which provides excitation energy. On the other hand, the excited cavity supplies energy to the sample of interest (loading) through coupling [10,11]. The basic structure of a cavity set is sketched as in Fig.2(a), which consists of a wave guide, a coupler with iris control, and a metallic cavity. The working principle of coupled cavity can be understood by using LRC equivalent circuit which couples to a transmission line [10,11] as shown in Fig.2(b) and 2(c). The coupling structure is represented by an ideal transformer with transforming ratio 1:n. The parallel RLC circuit could be transformed to T_s from AB plane via the transformer, such that $C'=n^2C$, $R'=R/n^2$, and $L'=L/n^2$. The resonance frequency, f_{res}, and Q-factor will not be affected by the transformation [10] as pointed out by eq.(1) and eq.(2), respectively:

$$\omega_0' = \frac{1}{\sqrt{L'C'}} = \frac{1}{\sqrt{\frac{L}{n^2}n^2C}} = \omega_0 \tag{1}$$

$$Q_0' = \omega_0'C'R' = \omega_0 n^2 C \frac{R}{n^2} = \omega_0 CR = Q_0 \tag{2}$$

A coupling coefficient, β, which defines the relation between energy dissipation (E_d) in a cavity and energy dissipation of external circuit (E_e), tells if the cavity is coupled with external circuit.

$$\beta \equiv \frac{E_d}{E_e} = \frac{\frac{U^2}{2Z_C}}{\frac{U^2}{2R/n^2}} = \frac{R}{Z_C n^2} = \frac{R'}{Z_C} \tag{3}$$

U in eq.(3) stands for the voltage applied to the cavity, and 1:n is the transforming ratio for tuning to critical couple state ($Z_C = R' \Rightarrow \beta = 1$) before doing FMR measurement. The characterization impedance, Z_C, is always designed to be 50Ω (Z_0) for matching impedance as mentioned previously. Signal will be distorted and weakened if working at either under-coupling (β<1) or over-coupling states (β>1). The external Q factor (Q_e) of a cavity, coupled to external circuit, states the energy dissipation in the external circuit, and also has the following relationship with β:

$$\beta = \frac{Q_0}{Q_e} \tag{4}$$

In eq.(4), Q_0 is the intrinsic Q factor of cavity. Practically, the loaded Q-factor, Q_L, is used in experiments:

$$Q_L = Q_0 \frac{\beta}{1+\beta} \text{ ; or } \frac{1}{Q_L} = \frac{1}{Q_0} + \frac{1}{Q_e} \tag{5}$$

Metallic cavities used for FMR measurement have to be TE mode, such that the cavity center which is also the sample location has maximum excitation magnetic field. Furthermore, this kind of cavity has very high Q-factor and plays key role in FMR detection. The signal output, V_s, from the cavity can be expressed as [15]:

$$V_S = \chi'' \eta Q_L \sqrt{PZ_0} \qquad (6)$$

χ'', η, P and Z_0 in eq.(6) stand for imaginary part of magnetic susceptibility of the sample, filling factor, microwave power, and characteristic impedance, respectively. The Q-factor of a cavity directly relates to the detecting sensitivity, which depends very much on the design and manufacturing technique. A good metallic cavity normally has unloaded Q-factor of more than 5,000 which cannot be changed after being manufactured, and therefore is regarded as a constant in eq.(6). Z_0 of 50Ω is fixed for matching the network impedance. We cannot play too much on χ'' and η as well, since the former is the intrinsic behaviour of the sample of interest to be tested, and the last parameter is the volume ratio of the sample and cavity. For thin film and multilayer samples, η is very small, hence this term cannot contribute too much to V_S. The source power, P, at the first glance, is the only possible adjustable parameter for sensitivity. Due to the occurrence of saturation, higher power may not be that helpful in increasing the FMR signal. Large excitation power drives magnetization precession in non-linear region which complicates the spectrum analysis. Besides, signal will be reduced and broadened while operating at saturation region. In order to determine the line shapes and widths precisely, this region should be avoided, hence low power is a good choice. The determination of saturation power is not difficult, since the signal intensity grows as the square root of power as indicated by eq.(6). Checking the signal intensity to see if eq.(6) is still validated as the power is increased. It is noted that the maximum excitation power of a common VNA is just a few mWs, and V_S is not that clear to follow $\sqrt{P}$.

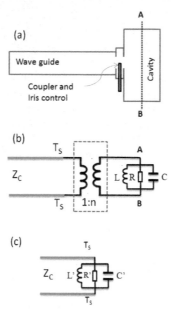

Figure 2. (a) The layout of a resonant cavity, (b) The equivalent LRC circuit. The iris control is regarded as a transformer, and (c) The equivalent circuit transformed to the T_s plane [10].

The Q-factor of commercial or home-made cavities can be determined easily by using a VNA, which can also be expressed as:

$$Q = \frac{f_{res}}{\Delta f} \tag{7}$$

f_{res} is the resonant frequency of a cavity, and Δf, is the full width at half maximum (FWHM) of the resonant peak. Eq.(7) can be simply found by a built-in function of VNA which measures directly the reflected power of the cavity in dB at f_{res}. Besides with dB unit, FWHM (Δf) is also obtained at half power point, i.e. the power level at -3dB. As the values of f_{res} and FWHM are known, Q is determined as shown in Fig.3(d).

Q will be decreased once a sample is inserted. This is because the inserted sample and its holder absorb energy and change the coupling conditions resulting in Q reduction and resonance frequency shift. However, these deviations can be amended a little bit by adjusting the iris which controls the effective impedance by varying the aperture size between the cavity and wave guide, and by adjusting the position of metallic cap. This is equivalent to change the transformer ratio, that is, the equivalent inductance and capacitance. The iris control is a plastic screw (low μ-wave absorption material) which has a gold coated metal cap on one end, and the effective impedance of the whole (wave guide + iris + cavity) depends on the size of the aperture and location of the metal cap. The function of the iris acts as a device to tune the L and C, such that the cavity is critical couple again. As field is swept, the variations of sample's magnetization break the coupling condition so the cavity is deviated from critically coupled state. Consequently, wave is reflected back to the ⊚-wave bridge, and FMR signal is resulted.

For an ideal metallic cavity (infinite conductivity and perfectly smooth inside wall surface), there will be a unique resonant peak with infinite large Q-factor, However, this is not the case in practice. Since the cavity itself has finite conductivity, and further, the inside wall is not perfectly flat, these cause the existence of many peaks with very low Q values as extracted by a VNA shown in Fig.3(a). Fig.3(b) shows the simulation of a copper made cavity with a very rough inside wall. As the roughness is reduced, the spectrum is clearer as shown in Fig. 3(c). This could be due to the μ-wave which is multi-scattered by lumps on the wall surface, and these lumps could also enhance power dissipation further. In consequence, there exists many low Q peaks. Even there are many peaks other than the eigen frequency, these peaks do not response to the change of magnetization, and hence useless for FMR characterization. The manufacture of metallic cavity with Q-factor of higher than 3,000 is laboring. This is because the inside wall has to be mirror polished and coated with a few μm thick Au layer to reduce the imperfection.

Fig.3(a) and Fig.3(d) are the frequency response of a Bruker X-band cavity (TE$_{102}$, ER 4104OR) exhibiting in larger and smaller frequency windows, respectively. The unloaded Q-factor of this cavity is ~9,000 as claimed by the manufacturer, however, our measurement tells that the Q factor is more than 14,000. This is because frequency resolution of the VNA is very high. As the resolution is decreased, Q is found to be decreased as well.

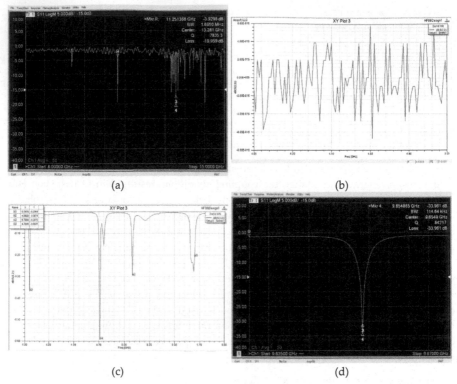

(a)   (b)

(c)   (d)

Figure 3. (a) the frequency response of a X-band cavity. Simulation result of a Cu cavity with (b) a very rough inside wall surface, and (c) a smooth inside wall surface. (d) VNA measurement of an X-band cavity with Q higher than 87, 000. Note that there are many side peaks but not appears in this frequency range

2.2. Shorted waveguide cavity

The fabrication of metallic cavity with very high Q-factor is not easy as mentioned above. If the lossy of the sample is not very big, cavity with Q of a few hundred to thousand may be good enough to recognize FMR. If so, a shorted waveguide cavity (SWC) could be an alternative. This kind of cavity is just a section of waveguide sealed by a metal plate at one end [16] as show in Fig.4. Since metallic wave guide are commercially standard with wide range of frequency from L to W band, it is worth to obtain FMR information at different bands with this cheap, simple and effective method. If the length of the waveguide tube is equal to $n\lambda/2$, standing wave can be formed inside. In that n and λ represents for integer and wave length of the μ-wave, respectively. In order to excite FMR signal, sample has to be placed at the $n\lambda/4$ position away from the shorted plate at which maximum magnetic field is located. In Fig.4, n=0 and n=1 are the suitable locations, however, n =0 gives advantages of easier sample manipulation, and lesser interference from irrelevant insertions. Sample is mounted at the top of a plastic screw through the shorted plate, such that its position can be adjusted for maximum signal. However, SWC does not have an iris to tune the impedance, hence critical coupling may not be easy to obtain. Although the Q is somewhat lower (less than 2,000), SWC is easy to make from commercial waveguide tube. Furthermore, different band frequency cavity can be built in the same manner, and this is particular convenient to study FMR at different band frequency with network analyzer.

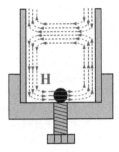

Figure 4. A shorted wave guide with possibility of sample manipulation. Magnetic field distribution originated from the μ-wave is also sketched [16].

2.3. Microstrip and co-planar waveguide

Microstrip (MS) and co-planar waveguide (CPW) which are indeed transmission lines, are commonly used to extract FMR at broad frequency range [16,17,18,19]. Sample is mounted on top of their signal lines, and μ-wave is conducted into MS (CPW) via Port 1 of the VNA, and the differential change in either reflected or transmitted signal is analyzed. In this operation scheme, frequency is swept at fixed field, and once the FMR conditions are fulfilled μ-wave absorption occurs. Fig.5 and Fig.6 are the sketches of MS and CPW. Since h << λ, μ-wave propagates in these two lines is regarded as quasi-TEM mode. In order to match the impedance of 50Ω, the dimensions of these two planar transmission lines are restricted.

In case of MS, we have the followings [11]:

The effective dielectric constant, ε_eff has approximately the form:

$$\varepsilon_{eff} = \frac{\varepsilon_r+1}{2} + \frac{\varepsilon_r-1}{2}\frac{1}{\sqrt{1+12h/w}} \qquad (8)$$

The dimension of the strip line and characteristic impedance, Z_C, can be worked out as below:

$$Z_C = \begin{cases} \dfrac{60}{\sqrt{\varepsilon_{eff}}} \ln\left(\dfrac{8h}{w} + \dfrac{w}{4h}\right) & ; \text{ for } w/d \le 1 \\[4mm] \dfrac{120\pi}{\sqrt{\varepsilon_{eff}}\left[\dfrac{w}{d}+1.393+0.667\ ln\left(\dfrac{w}{d}+1.444\right)\right]} & ; \text{ for } w/d \ge 1 \end{cases} \qquad (9)$$

In case of CPW if the thickness of signal line is ignorable, then [10]:

$$\varepsilon_{eff} = \frac{\varepsilon_r + 1}{2} \tag{10}$$

$$Z_C = \begin{cases} \dfrac{\eta_0}{\pi\sqrt{\varepsilon_{eff}}} \ln\left(2\sqrt{\dfrac{b}{w}}\right) \ \Omega & ;\ 0 < w/b < 0.173 \\[2em] \dfrac{\pi\,\eta_0}{4\sqrt{\varepsilon_{eff}}}\left[\ln\left(2\dfrac{1+\sqrt{w/b}}{1-\sqrt{w/b}}\right)\right]^{-1} \ \Omega & ;\ 0.173 < \dfrac{w}{b} < 1 \end{cases} \tag{11}$$

η_0 in eq.(11) stands for the wave impedance in free space. The Z_C = Z_0 is set to 50Ω, then b, d and w can easily be determined from eq.(9) and eq.(11). The line width of the strip can be ranged from sub-millimeter to millimeters depending on the design and dielectric material used as implied by these equations. The size of this kind of line width can be fabricated simply by photo lithography together with life-off process. Due to the large loss, the characterization of nano size sample may not be easy and lock-in technique is needed for SNR enhancement as described in next section.

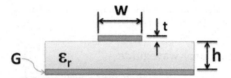

Figure 5. The geometry of a microstrip line, where εr is the relative dielectric constant of the substrate.

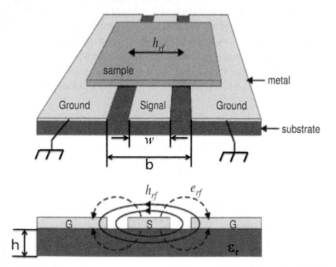

Figure 6. Schematical diagram of a coplanar waveguide along with the dimension. The lower diagram shows the magnetic (full lines) and electric (dashed lines) field lines winding around the signal conductor (S).

FMR can determined by characterizing the transmittance, ie. S_{12}. However, if shorted MS, and CPW are used, S_{11}, the reflectance FMR can also be found. Fig.7 shows the FMR of permalloy films extracted by VNA and CPW [18].

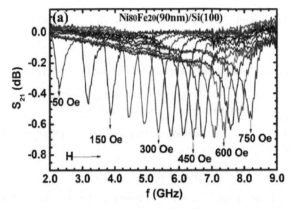

Figure 7. A typical FMR transmittance spectrum of permalloy films extracted by VNA with CPW [18].

3. Technique for signal to noise enhancement

3.1. Lock-in phase detection

If signal is deeply buried inside noise floor, lock-in method is normally employed to enhance the signal to noise ratio (SNR) by narrowing the detecting band window [20]. Considering a DC Signal, S, which is heavily contaminated by white noise, n, and for every measurement comes out from a measurement equipment, a voltmeter for an example, the reading can be expressed as:

$$V = S + n \tag{12}$$

If S can be modulated sinusoidally at a certain frequency somehow, then:

$$V(t) = S\cos(\omega t) + n \tag{13}$$

Note in eq.(13) that the modulation of white noise is still a white noise.

There is also a reference signal with same frequency but could be different in modulating amplitude (V_a) and phase (ϕ), i.e.,

$$V_{ref} = V_a \cos(\omega t + \phi) \tag{14}$$

The product of eq.(13) and (14) gives:

$$V(t)V_{ref} = SV_a \cos(\omega t)\cos(\omega t + \phi) + nV_a \cos(\omega t + \phi) =$$
$$\frac{SV_a}{2}[\cos(\phi) + \cos(2\omega t + \phi)] + n\, V_a \cos(\omega t + \phi) \tag{15}$$

The second and third terms can be eliminated if these two parts are passed to a low pass filter with cutoff frequency setting at $\omega/2$ or lower. Finally, we have:

$$V \propto \frac{SV_a}{2}\cos(\phi) \tag{16}$$

That is, the signal is amplified by V_a, and has a maximum while ϕ is "phase-locked" to zero, hence the name of lock-in amplifier.

3.2. Feld modulation lock-in technique and derivative spectrum

It is easier to determine H_{res}, and ΔH_{pp} by differentiating the original signal. To do so, field modulation lock-in detection is employed and described below.

There is a small field, H_a, with modulation frequency of ω superposing on top of external DC magnetic field, H_0:

$$H(t) = H_0 + H_a \cos(\omega t) \tag{17}$$

Assuming we have FMR signal, V_{FMR}, whose Taylor expansion at H_0 is given below:

$$V_{FMR}(H) = V_{FMR}(H_0) + \frac{dV_{FMR}}{dH}\Big|_{H=H_0} H_a \cos(\omega t) + \cdots \tag{18}$$

Meanwhile, we also have another signal, the reference, which has the same frequency as that of the modulation field, but with a phase angle, ϕ:

$$V_{ref} = \cos(\omega t + \phi) \tag{19}$$

The product of eq.(18) and (19) gives eq.(20) below:

$$V_{FMR}(H) \times V_{ref} = V_{FMR}(H_0)\cos(\omega t + \phi)\frac{dV_{FMR}}{dH}\Big|_{H=H_0} + H_a\cos(\omega t)\cos(\omega t + \phi) + \cdots =$$
$$V_{FMR}(H_0)\cos(\omega t + \phi) + \frac{1}{2}\frac{dV_{FMR}}{dH}H_a\cos(\phi) + \frac{1}{2}\frac{dV_{FMR}}{dH}H_a\cos(2\omega t + \phi) + \cdots \tag{20}$$

The first and third terms in eq.(20) can again be removed by using a low pass filter with cutoff frequency setting at $\omega/2$ or even lower. The second term is time independent, proportional to derivative of the input signal and magnify by H_a, which has a maximum if the phase, ϕ, is locked to 0. Another advantage of use this method is that $1/f$ noise and drift problem can be excluded by setting the sampling window at high frequency.

The way to turn the DC or quasi DC signal from to AC is an important subject. The simple and normal way to do this is to insert two coils which sandwich the cavity. These coils are driven by a power amplify at certain high frequency, such that an AC magnetic field of a few mT at a frequency up to 200 kHz can be produced. Therefore, the output signal consists of the modulation and DC components. All these together with the reference are sent to the lock-in amplify for SNR enhancement and derivative spectrum as mentioned previously. Modulation amplitude (MA), modulation frequency (MF), and time constant (TC) which is the reciprocal of the low pass filter's cutoff frequency, have large influence on the spectra.

Signal strength is linearly proportional to the MA while MA is small. Simply speaking, MA should be small enough for sampling in linear region, but needs to be large enough for gaining good sensitivity. However, too large the MA results in signal distortion as shown in Fig.8(a). This can be understood from eq.(20) that high order term cannot be ignored for large MA which causes distortion of derivative signal. The choice of MF is critical as well that small MF cannot get rid of 1/f noise completely, and elongates the acquisition time. Large MA and MF also cause passage effect that the rate of "passage" through the absorption line is faster than the relaxation rates, which results in distorted spectrum, or even inversion of signals upon reversal of the field scan direction. Therefore, it is important to check the FMR signal with either a standard or a well-known sample with different lock-in conditions for the best settings. In order to obtain correct line shape spectrum, the rule of thumb is to set the MA about 1/10 of the FWHM and increases to about 1/3 if necessary. TC also needs to coordinate with MF. According to Nyquist sampling theorem, the sampling rate should be at least twice the highest frequency contained in the signal [21]. For example, if the MF is 100 kHz, the sampling frequency would be at least 200 kHz at which TC is about 5µS. If this restriction does not fulfill, sampling points cannot not be captured immediately. Consequently, information loss and line shape distortion are always resulted. The best way to avoid this is to set the scan time for a FMR signal 10 times longer than the time constant.

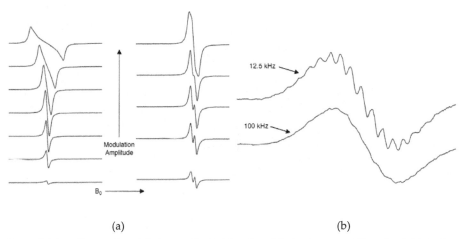

(a)   (b)

Figure 8. The influence of modulation amplitude (a) and modulation frequency (b) on signal. Distortion will be resulted if MA and MF are not appropriate [15].

4. Vector analyzer based FMR spectrometer

4.1. Basics of vector network analyzer (VNA)

VNA is indeed an instrument developed for characterization of electrical devices (device under test, DUT) by sending an electromagnetic wave. As an analogue of that in optics, the incident wave (either optical or micro wave) will be reflected or/and transmit after interacting with the DUT. By examining the reflectance and transmittance, that is, the ratios of the powers of reflected, transmitted to that of the incident waves, scattering parameters of the DUT can be found as depicted in Fig.9. VNA cannot only find out the reflectance and transmittance, but also the impedance, phase lag, insertion loss, return loss, voltage standing ratio, etc., can be worked out. However for FMR experiment, only the first two functions are employed. VNA has at least two ports, and each port can produce and measure μ-wave. The scattering parameter S_{ij} stands for scattering power ratio of incident wave produced by port i, and measured by port j. That is, for a two-port VNA, S_{11} and S_{22} characterize wave reflectance, while S_{12} and S_{21} determine the transmittance. If FMR is determined by reflected spectrum, one port is enough by analyzing either the $S11$ or $S22$. For transmittance spectrum, two ports are required for either the S_{12}, or S_{21} determination. VNA has capability of microwave generation and detection, and furthermore, nowadays model has frequency ranged from about 0.1 to 100 GHz, or even higher at excitation power of tens dBm, Thus, it can serve as a microwave bridge to extract FMR parameters at a very broad frequency band. VNA measures not only the scattering amplitudes, but also their correlated phases. This is in contrast to its counterpart, the scalar network analyzer (SNA) which cannot tell phase information. FMR signal is extracted from power absorption which contains no phase information, and SNA should be good enough for the purpose. However, this kind of machine has been obsoleted for many years.

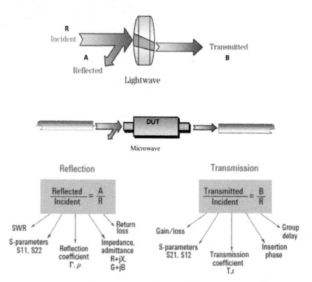

Figure 9. The fundamentals of microwave network analysis are analogue to that of the optics. The lower part of the panel indicates the definitions of scattering parameters [24].

VNA is commonly used with MS and CPW for FMR determination with frequency swept at fixed field. Since signal output from VNA cannot be plugged into lock-in amplifier directly, the characterization of nano scale sample is thus difficult without using lock-in. The application lock-in with VNA will be discussed in next section.

4.2. Field swept VNA-FMR

Metallic cavity usually has rather higher Q-factor and could be simply used with a VNA for field swept FMR detection. Taking the advantages of high Q cavity and VNA, C.K. Lo, *et. al.* [23] demonstrated FMR measurement without employing field modulation lock-in technique. They employed a TE_{102} cavity with unload Q of ~9,000 at X-band, and FMR signal was recorded through the measurement of reflectance, and therefore one VNA port for S_{11} was used. This powerful method extracts signal easily and directly as shown in Fig.10(a). One could if necessary, differentiate the original data for derivative spectrum which is used to determine the peak position and line wide as those in commercial FMR spectrometer. Further, this combination has quite good sensitivity that 1.6 nm CoFeB can be detected with SNR better than 5 as shown in Fig.10(b).

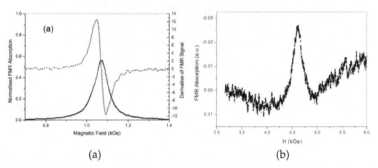

(a)   (b)

Figure 10. (a) FMR of 5 nm Py exacted at P = 10 dBm. The dotted line is the derivative of experimental data. (b) Signal extracted from a 1.6nm CoFeB

A new built spectrometer should be characterized before properly used and well known samples, such as permalloy, Fe, Co, etc., ferromagnetics are normally employed due to their ΔH and peak positions can be found widely in literatures. Also, these samples are easy to prepare with different thicknesses, and results should be comparable to those in literatures reported.

There are many build-in useful functions with nowadays VNA, and only some of them are used, for instances, the traces of valley (dip) position, Q factor, band width, average, etc. Dip frequency and Q-factor will be varied as the sample's magnetization state is changed by external field. The changes of these quantities were also recorded simultaneously for a Fe/Ag multilayer as shown in Fig.11 in which (a) is the FMR absorption, and (b), the post derivative of (a). The shifts of resonant frequency and Q-factor are shown in (c) and (d), respectively. Despite the variation of center frequency is just a few hundreds MHz, it allows us to determine ferromagnetic parameters precisely. The change of Q-factor is upside-down to that of the absorption. This is because any absorption of microwave inside the cavity will reduce the Q value.

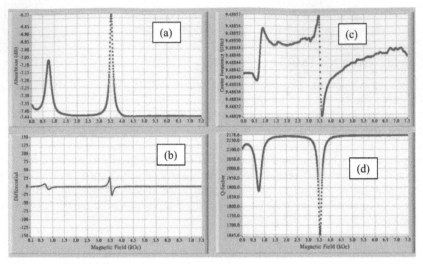

Figure 11. The changes of FMR (a), resonance frequency (c), and Q-factor are recorded. (b) is the derivative spectrum taken from (a) mathematically.

The excitation power of VNA is only tens dBm, and it is unlikely to drive the sample into non-linear regime. Signal intensity is proportional to $\sqrt{P}$ as indicated by eq. (6), however, SNR is found to be not much different if P is bigger than -20 dBm as demonstrated in Fig.12.

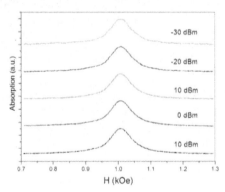

Figure 12. FMR signal of 10nm Py was recorded as function of excitation power. Note that these spectra have the same scale, but different offset for clear comparison.

As mentioned previously that VNA has very high frequency resolution, apparently, Q could be tuned as high as 140,000 and more by simply adjusting the iris. However, distorted spectrum is resulted at very high Q, and two peaks appear while Q is adjusted close to maximum Q as seen in Fig.13(a) and (b), respectively. It is also found that the SNR does not varied significantly if Q lies between 2,000 to 13,000 (the unloaded Q is about 9,000 as claimed). Surely, this finding is just for referral, and would depend very much on cavity used.

The advantages of the VNA-FMR with high Q cavity are: (1) the sensitivity which is comparable to that of commercials, but a lot cheaper, (2) multi frequencies can be done in

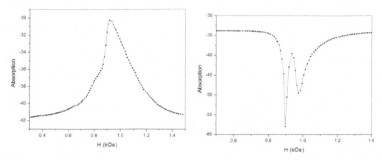

Figure 13. (a) data taken at Q ≈ 50,000, and (b) at ~110,000. Clearly, distorted signals are obtained.

the same manner by just changing to another band cavity only within the frequency range of the VNA. Nevertheless, some points needed to be well aware. Firstly, VNA has built-in sweep average function which increases the SNR for every sweeping. The penalty of using high average count is the very long acquisition time. For example, a commercial FMR spectrometer takes about a few seconds to sweep a spectrum over 2 kOe field at 1 Oe resolution with time constant of a 1mS. However, the settled time for a VNA with average of 1 is found to be about 0.1S a point. Therefore, it takes at least 3 min to obtain the spectrum with the same conditions. Average of 10, and sometime 50 are often used for better SNR, then the acquisition time is very long which could overheat the electromagnet. Secondary, VNA cannot work with lock-in amplifier directly, and a circulator is also needed to force the reflected wave to a new port. Before the signal is conducted to the lock-in amplifier, a microwave transducer (Schottky diode, for example) is required. This is because the output from VNA is power which has to be transduced into voltage before plugging into the lock-in amplifier. Besides, the output of VNA has already been averaged over a range of frequencies. Thus the Schottky detector cannot recognize this tiny average change. This problem can be amended by narrowing the starting and ending frequencies of the VNA. The two frequencies indeed can be set to the same value as the dip frequency. While working with lock-in, the VNA-FMR acquires data as fast as the commercial one, since the SNR enhancement is processed by the lock-in and not by the VNA. Further, MS-, and CPW- FMR can also work together with lock-in method in this manner. If one wants to study high power VNA-FMR, a high power microwave amplifier could be used. However, if the VNA is still used as the analyzer, a suitable attenuator has to be inserted between the returned line and VNA, otherwise damage is resulted.

Author details

Chi-Kuen Lo
*Department of Physics, National Taiwan Normal University, Taipei,
Taiwan*

5. References

[1] M. Diʹaz de Sihues, C. A. Durante-Rincon, and J. R. Fermin, J. Magn. Magn. Mater. 316, e462 (2007).

[2] V. G. Gavriljuk, A. Dobrinsky, B. D. Shanina, and S. P. Kolesnik, J. Phys. Condens. Matter 18, 7613 (2006).

[3] Z. Zhang, L. Zhou, P. E. Wigen, and K. Ounadjela, Phys. Rev. B 50(9), 6094 (1994).

[4] B. Heinrich, Y. Tserkovnyak, G. Woltersdorf, A. Brataas, R. Urban, and G. E. W. Bauer, Phys. Rev. Lett. 90(18), 187601 (2003).

[5] V. P. Nascimento and E. Baggio Saitovitch, F. Pelegrinia, L. C. Figueiredo, A. Biondo, and E. C. Passamani, J. Appl. Phys. 99, 08C108 (2006).

[6] K. Lenz, E. Kosubek, T. Tolinski, J. Lindner, and K. Baberschke, J. Phys.: Condens. Matter 15, 7175 (2003).

[7] M. Oogane, T. Wakitani, S. Yakata, R. Yilgin, Y. Ando, A. Sakuma, T. Miyazaki, J. J. Appl. Phys. 50 (5A) 3889 (2006)

[8] R. Urban, G. Woltersdorf, and B. Heinrich, Phys. Rev. Lett. 87(21), 271204-1 (2009).

[9] T. Kato, K. Nakazawa, R. Komiya, N. Nishizawa, S. Tsunashima, and S. Iwata, IEEE Trans. Magn., VOL. 44, NO. 11, NOVEMBER 2008

[10] "Microwave Electronics: Measurement and Materials Characterization", L. F. Chen, C. K. Ong, C. P. Neo, John Wileys & Sons Ltd. 2004

[11] "Microwave Engineering", D.M. Pozar, John Wileys & Sons Ltd. (2012)

[12] "High-Frequency EPR Instrumentation", E.J. Reijerse, Appl Magn Reson (2010) 37:795–818

[13] See http://www.bruker-biospin.com/epr-products.html.

[14] See http://www.jeol.com/PRODUCTS/AnalyticalInstruments/Electron SpinResonance/tabid/98/Default.aspx.

[15] "Quantitative EPR", G. R. Eaton, S. S. Eaton, D. P. Barr, and R. T. Weber, Springer Wien, New York, (2010)

[16] Marina Vroubel, Yan Zhuang, Behzad Rejaei, and Joachim N. Burghartz, J. Appl. Phys. 99, 08P506 (2006)

[17] C. Nistor, K. Sun, Z. Wang, M. Wu, C. Mathieu, and Matthew Hadley Appl. Phys. Lett. 95, 012504 (2009)

[18] Y. Chen, D.S. Hung, Y.D. Yao, S.F. Lee, H.O. Ji, C. Yu, J. Appl. Phys. 101, 09C104 (2007)

[19] H. G LOWI'NSKI, J. DUBOWIK, Acta Physicae Superficierum , Vol. XII, 2012

[20] "The Art of Electronics", P. Horowitz, W. Hill, Cambridge Univ. Press, 2nd Ed. 1989

[21] Advanced Digital Signal Processing and Noise Reduction", S.V. Vaseghi, John Wiley & Sons Ltd., 2008

[22] Agilent VNA user manual

[23] C.K. Lo, W.C. Lai, J.C. Cheng, Rev. Sci. Instruments, 82, 086114 (2011)

[24] R.W. Damon, Rev. Mod. Phys., 25, No.1, 239 (1953)

[25] N. Bloembergen, S. Wang, Phys. Rev., 93 No.1 72 (1953)

Chapter 5

Detection of Magnetic Transitions by Means of Ferromagnetic Resonance and Microwave Absorption Techniques

H. Montiel and G. Alvarez

Additional information is available at the end of the chapter

1. Introduction

Due to nature of the magnetic materials, the experimental techniques to study their physical properties are generally sophisticated and expensive; and several techniques are used to obtain reliable information on the magnetic properties of these materials. One of the most employed techniques to characterize the magnetic materials is the electron magnetic resonance (EMR), also well-known as the ferromagnetic resonance (FMR) at temperatures below Curie temperature (T_c) and the electron paramagnetic resonance at temperatures above T_c. EMR is a powerful technique for studying the spin structure and magnetic properties in bulk samples, thin films and nanoparticles, being mainly characterized by means of two parameters: the resonant field (H_{res}) and the linewidth (ΔH_{PP}); these parameters reveal vital information on magnetic nature of the materials (Montiel et al., 2004, 2006; Alvarez et al., 2008, 2010). It is also necessary to mention that EMR is one of the most commonly used techniques to research the dependence of the magnetic anisotropy with respect to orientation of the sample (Montiel et al., 2007, 2008; G. Alvarez et al., 2008) and the temperature (Montiel et al., 2004; Alvarez et al., 2010); this technique is also applied to study magnetic relaxation in solid materials through their linewidth, and that it is due to conduction mechanisms and intrinsic relaxation.

In particular, EMR technique is employed with success to determine the onset of magnetic transitions (Okamura, 1951; Okamura et al., 1951; Healy, 1952; Montiel et al., 2004; Alvarez et al., 2006, 2010), such as the Néel transition (from a paramagnetic phase to antiferromagnetic ordering) and the Curie transition (from a ferromagnetic or ferrimagnetic order to paramagnetic phase); through changes in the spectral parameters. EMR around critical temperatures have been object of an active research, e.g. Okamura et al. (1951) have studied the Néel transition in some materials by means of EMR technique. Fig. 1 shows the direct resonance absorption versus magnetic field for MnS in the 78-329 K temperature range. The height of the resonant absorption decreases and the linewidth is become broader with decreasing temperature, especially below the Néel temperature (T_N= 160 K); where the position of the maximum is found at a constant magnetic field of 3510 G for all the temperatures, suggesting the existence of a strong local magnetic field in this material.

The previous behavior is quite contrary to those of the ferromagnetic and ferrimagnetic materials around Curie transition. For example, Healy (1952) has employed EMR technique to study the nickel ferrite ($NiFe_2O_4$) in temperature range of 78 K to 861 K. In Fig. 2, the direct resonance spectra show that the linewidth decreases and a shift in resonant field were observed with increasing temperature, all these changes can be completely associated with the Curie transition (T_c= 858 K).

Okamura (1951) also carries out a characterization in the ferrite $NiFe_2O_4$ by means of EMR technique but at low temperature. Fig. 3 shows the direct resonance absorption with varying applied dc magnetic field at various temperatures in a disk $NiFe_2O_4$. In these spectra a double absorption were clearly observed, and both absorptions are dependent of the temperature; where one of the absorptions is near to zero magnetic field. In our EMR measurements, we also detected a second absorption mode around zero field in amorphous alloys (Montiel et al., 2005) and ferrites (Montiel et al., 2004; Alvarez et al., 2010); however, a detailed discussion on this absorption type is gathered to continuation.

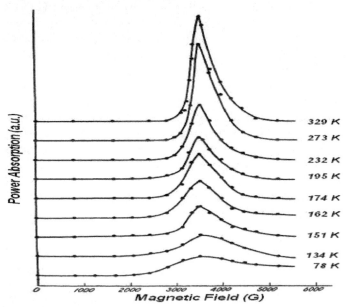

Figure 1. Direct resonance absorption versus dc magnetic field at 9.3 GHz for MnS around Néel transition (adapted from Okamura et al., 1951).

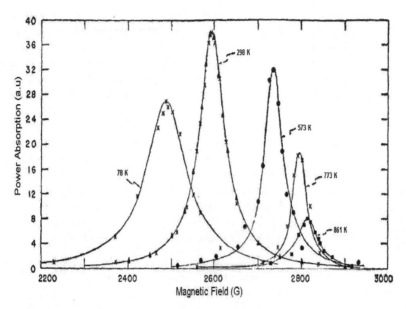

Figure 2. Direct resonance absorption versus dc magnetic field at 9 GHz in a sphere NiFe$_2$O$_4$ around Curie transition (adapted from Healy, 1952).

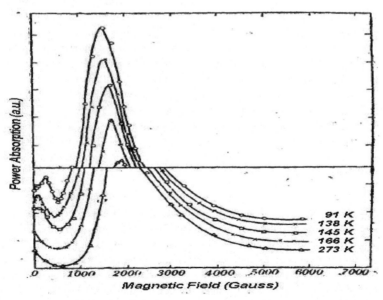

Figure 3. Direct resonance absorption versus dc magnetic field at 9.3 GHz for a disk NiFe$_2$O$_4$ at low temperature (adapted from Okamura, 1951).

The non-resonant microwave absorption (NRMA) was used in 1987 to detect the transition between the normal state and superconducting state in high-T$_c$ superconductor ceramics (Bhat et al., 1987; Blazey et al., 1987; Bohandy et al., 1987; Khachaturyan et al., 1987; Moorjani et al., 1987). This was followed by a large number of reports on not only high-T$_c$ superconductor ceramics (Kim et al., 1993; Topacli, 1996, 1998; Velter-Stefanescu et al., 1998, 2005; Padam et al., 1999, 2010; Shaltiel et al., 2001; Alvarez & Zamorano, 2004), but also including organic superconductors (Zakhidov et al. 1991; Bele et al., 1994; Hirotake et al., 1997; Niebling et al., 1998; Stankowski et al., 2004), the conventional superconductors of type-I and type-II (Kheifets et al., 1990; Bhide et al., 2001; Owens et al., 2001; Andrzejewski et al., 2004) and the newly discovered iron pniticide (Panarina et al., 2010; Pascher et al., 2010). Researches on NRMA have shown that this phenomenon is highly sensitive to detection of a superconducting phase in a material under study. The NRMA is usually detected as a function of a dc applied magnetic field or temperature, where these two variants are historically known as the magnetically-modulated microwave absorption spectroscopy (MAMMAS) and the low-field microwave absorption (LFMA), respectively.

In Fig. 4(a) is shown the MAMMAS response for bulk sample of the ceramic superconductor Bi-Sr-Ca-Cu-O around the superconducting transition. This response shows a level of absorption constant from 300 K to T$_{on}$= 81.5 K, and suddenly this signal rises sharply until T$_{max}$=63.6 K; at that temperature the superconducting transition has been completed. As temperature goes down from T$_{max}$, the superconductor sample enters more and more into the mixed state with a rigid fluxon lattice, and a decrease in the microwave absorption is observed. Fig. 4(b) shows LFMA spectra for selected temperatures, in a bulk sample of the ceramic superconductor Bi-Sr-Ca-Cu-O, where a hysteresis loop is observed in the superconductive state; since the microwave induced dynamics of the fluxon is dissipative, the field sweep cycle in LFMA measurement shows a hysteresis. The physical meaning of this hysteresis has been amply discussed by Topacli (1998) and Padam et al. (1999). For T$\geq$ 63 K, a non-hysteretic LFMA signal is observed and which goes disappearing when increasing the temperature, i.e. when the sample enters to normal state. The above-mentioned is a striking example of that the transition to the superconductive state in superconductor ceramics leads to NRMA. Additionally, it is recognized that NRMA is due mainly to the dissipative dynamics within Josephson junctions and/or to induced currents through weak links in superconductor materials.

Today, it is safe to assume that all superconductor materials exhibit a NRMA, and which has been experimentally confirmed; but the reverse statement, that any material that exhibits a NRMA is a superconductor, in general is not true. The NRMA may be caused not only by superconductivity, but also by any phenomena associated with magnetic-field-dependent microwave losses in the materials, and it can be employed to detect the magnetic transitions (Nabereznykh & Tsindlekht, 1982; Owens, 1997; Alvarez & Zamorano, 2004; Montiel et al., 2004; Alvarez et al., 2007, 2009, 2010). Some antecedent studies are given in the following. Nabereznykh & Tsindlekht (1982) have reported a study of NRMA in nickel near the Curie transition (Fig. 5), and in particular, they have employed LFMA measurements for detecting

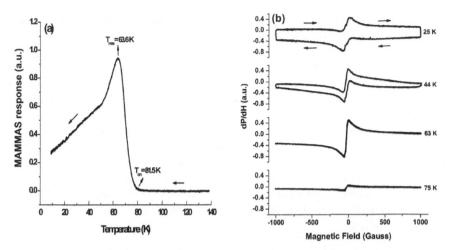

Figure 4. (a) MAMMAS response and (b) the LFMA spectra of a bulk sample of the ceramic superconductor Bi-Sr-Ca-Cu-O in the region of the superconducting transition.

the magnetic transition from a ferromagnetic order to paramagnetic phase. They suggest that the LFMA signal is due to the presence of a domains structure in the magnetic material. LFMA signal is highly distorted at T=571 K and at T=587 K it is completely inverted. At T= 601 K, the line intensity reaches its maximum, see Fig. 5; and when increases the temperature their intensity and width (ΔH_{LFMA}) diminishes. For T$\geq$ 630.9 K, a LFMA line is observed and it agrees with the value of the Curie temperature given in the literature for nickel, $\cong$ 630 K; suggesting that ΔH_{LFMA} is determined by the magnetic anisotropy field and the demagnetizing field.

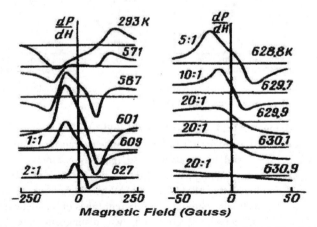

Figure 5. LFMA spectra in nickel for selected temperatures around Curie transition (adapted from Nabereznykh & Tsindlekht, 1982).

Other antecedent study is the detection of the Curie transition in a material with colossal magneto-resistance (CMR), as is $La_{0.7}Sr_{0.3}MnO_3$ manganite (Owens, 1997). The presence of the NRMA is evidence of the existence of a ferromagnetic order, i.e. this signal is not present in the paramagnetic phase and emerges as the temperature is decreased below Curie temperature, see Fig. 6; providing a sensitive detector of ferromagnetism. In Fig. 6(a), MAMMAS response shows the appearance and the rapid increase of the microwave absorption at ferromagnetic transition. Additionally, the half LFMA spectrum at 144 K in the ferromagnetic phase of $La_{0.7}Sr_{0.3}MnO_3$ is shown in Fig. 6(b). Owens (1997) suggests as a possible explanation of the origin of LFMA signal, the fact that the permeability in the ferromagnetic phase at constant temperature depends on the applied magnetic field, increasing at low fields to a maximum and then decreasing. Since the surface resistance of the material depends on the square root of the permeability, the microwave absorption depends non-linearly on the strength of the dc magnetic field, resulting in a NRMA centered at zero magnetic field.

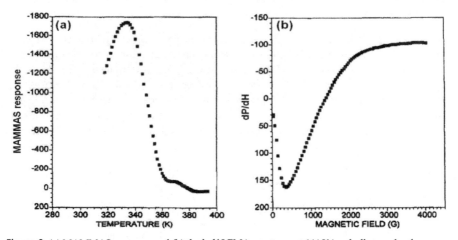

Figure 6. (a) MAMMAS response and (b) the half LFMA spectrum at 144 K in a bulk sample of $La_{0.7}Sr_{0.3}MnO_3$ manganite (adapted from Owens, 1997).

In this chapter, the changes in the EMR lineshape are studied for diverse magnetic materials in the 77-500 K temperature range; the different magnetic transitions are quantified by means of linewidth (ΔH_{PP}) and the resonant field (H_{res}) as a function of temperature. Through these studies we can distinguish the kind of present magnetic transition in the materials. Also, we employed the LFMA and MAMMAS techniques to give a further knowledge on magnetic materials, studying the different types of magnetic transitions and showing their main characteristics highlighted; we distinguish distinctive features associated with the microwave absorption by the magnetic moments and discuss on usefulness of these techniques as powerful characterization methodologies.

2. Experimental methods

In this work, the resonant and non-resonant microwave absorptions in several magnetic materials are studied around magnetic transitions. EMR technique measures the resonant microwave absorption as a function of dc magnetic field. Additionally, the NRMA measurements as a temperature function or an applied dc magnetic field are experimentally denominated as MAMMAS and LFMA, respectively.

2.1. Resonant microwave absorption measurement (EMR technique)

The basic components of a standard EMR spectrometer are shown in Fig. 7. In a sample placed at center of a microwave cavity (see Fig. 8), an applied dc magnetic field is increased until that the energy difference between the spin-up and spin-down orientations, match the microwave frequency of the power supply; where a strong absorption is clearly detected. For this type of spectrometer, the derivative of microwave absorption (dP/dH) as a magnetic field function is plotted.

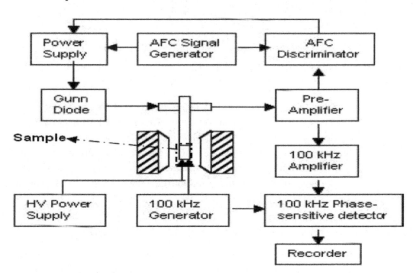

Figure 7. The diagram to blocks of a standard EMR spectrometer.

We give a description detailed of the EMR spectrometer (Jeol JES-RES3X). In a T-magic bridge, the microwaves (H_{mw}), with power P_0 and frequency v_{mw}= 8.8-9.8 GHz (X-band), are generated by a JEOL/ES-HX3 microwave unit and are fed to the cylindrical cavity (see Fig. 8) through a rectangular wave-guide; coupling adjuster of the cavity must be adjusted so that no wave is reflected from the cavity. The absorption of the microwave energy by a sample generates a change in the quality factor (Q) of the cavity; due to this change, the microwave bridge becomes non-balanced, causing that a wave is reflected from the cavity. The change in cavity Q-factor is due to changes in the energy absorption resonant by spins.

The reflected waves from the cavity (P_{ref}) with the information of the microwave absorption by the sample are directed towards a detector crystal, which was previously biased to a 10% of the incident power in order to work in the linear regime; with this method of detection-homodyne and lock-in amplification, a very high sensitivity in the measurements is guaranteed.

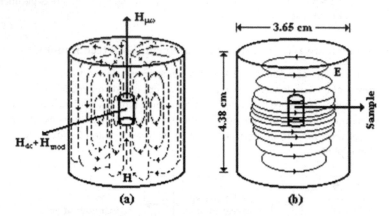

Figure 8. Distribution of (a) magnetic and (b) electric fields, and the sample location inside the TE_{011} cylindrical cavity in the JEOL JES-RES 3X spectrometer.

The sample is subjected to a dc magnetic field (H_{dc}), that it is produced by an electromagnet with truncated pole pieces, and a weak ac magnetic field (H_{mod}) is superimposed to H_{dc}. The H_{mod} is achieved by placing small Helmholtz-coils on each side of the cavity along the axis of the static field, which are fed and controlled by a sign generator. The amplitude of this field goes from 0.002 G to 20 G with a modulation frequency of 100 kHz, thus allowing, the microwave absorption registration at the modulation frequency. In EMR measurements, H_{dc} could be varied from 0 to 8000 G.

2.2. NRMA measurements (LFMA and MAMMAS techniques)

The Jeol JES-RES3X spectrometer was modified (see Fig. 9), connecting the output of a digital voltmeter (signal Y) to a PC enabling digital data acquisition (Alvarez & Zamorano, 2004); where this electrical signal is proportional to NRMA from sample. The signal Y is fed to a 7½ digits - Keithley DMM-196 voltmeter. Hence, the reading of this voltmeter (V_y) carries the information of the microwaves absorption by sample.

LFMA technique measures the NRMA as a function of H_{dc}, this uniform field is produced by the same electromagnet, but which receives current regulated from two power supplies (JEOL JES-RE3X and ES-ZCS2); and they are synchronized to obtain a true zero-value of the magnetic field between the pole caps. The Jeol ESZCS2 zero-cross sweep unit compensates digitally for any remanence in the electromagnet, with a standard deviation of the measured field of less than 0.2 G, allowing measurements to be carried out by cycling the H_{dc} about its zero-value continuously from -1000 G to +8000 G. Hence, symmetric field-sweeps from ±0.1 G to ±1000 G are available and asymmetric field-sweeps up to $-1000 \text{ G} \leq H_{dc} \leq 8000$ G are also available in order to detect possible hysteresis for NRMA signal, and which would point out to irreversible processes of microwave energy absorption. In this technique, the sample is zero field cooled or heated to the fixed temperature. For our studies, the temperature is maintained fixed with a maximum deviation of 1 K during the whole LFMA measurement (<8 min of sweeping). The magnetic field is swept following a cycle; the field sweep schemes have their analog in the magnetic hysteresis measurements. GPIB port of a PC receives the magnetic field coming from the Group3 DTM-141 teslameter, and it is displayed as the X-axis on the plot of the data being acquired; meanwhile, the voltmeter V_y receives the NRMA signal. Then, what is measured is not the microwave power absorption itself, but rather its derivative with respect to magnetic field (dP/dH). This allows us to distinguish the field-sensitive part of the microwave absorption from the part that does not depend on magnetic field, and record only the first; and also, to use narrow-band amplifier to enhance the signal, which greatly increases the signal-to-noise ratio.

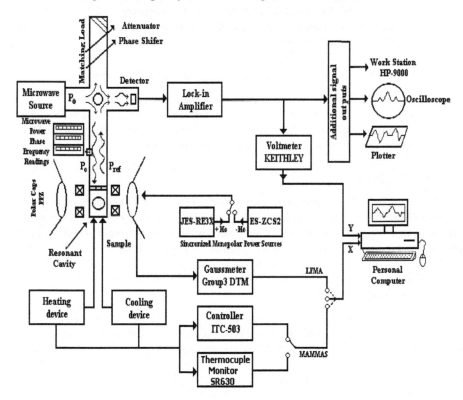

Figure 9. Block diagram of LFMA and MAMMAS techniques (adapted from Alvarez & Zamorano, 2004).

MAMMAS technique allows measurement as a temperature function, giving information on the temperature profile of NRMA response of each material and can also provide valuable information about the nature of magnetic ordering in the materials. The temperature of the sample is slowly varied (~1 K/min) and controlled by flowing N_2 gas through a double walled quartz tube, which is inserted through the center of the microwaves cavity, in the 77-500 K temperature range. The temperature is measured by a copper constantan thermocouple placed inside the sample tube just outside the cavity. Its output signal is further digitized by means of the Stanford SR630 thermocouple monitor. In this technique, the temperature is plotted along the X-axis, and the microwaves absorbed by the sample are collected as V_y and are plotted along Y-axis.

3. EMR and NRMA studies in magnetic transitions

In this section, we show several EMR and NRMA studies in diverse magnetic materials through magnetic transitions. These examples include: the amorphous ribbon $Co_{66}Fe_4B_{12}Si_{13}Nb_4Cu$, the ferrite $Ni_{0.35}Zn_{0.65}Fe_2O_4$ and the magnetoelectric $Pb(Fe_{2/3}W_{1/3})O_3$; highlighting their main characteristics and illustrating how magnetic transitions are manifested in this kind of measurements.

3.1. Curie transition (from a ferromagnetic order to paramagnetic phase)

We show several studies in amorphous ribbons of nominal composition $Co_{66}Fe_4B_{12}Si_{13}Nb_4Cu$ and dimensions of 2 mm wide and 22 μm thick, which were prepared by the melt-spinning method; where their initial amorphous state was checked by X-ray diffraction (XRD). All the measurements were performed in the 300-500 K temperature range.

3.1.1. EMR technique

These measurements were carried out from -1000 G to 8000 G, with forward and backward H_{dc} sweeps in order to detect reversible and/or irreversible microwave absorption processes. At 300 K, two microwave absorptions were observed: the first absorption to high magnetic field (~1469 G) corresponding to an EMR spectrum, and other absorption at low magnetic fields around zero (LFMA signal). Fig. 10 shows the derivative of microwave absorption, and which consists of an EMR spectrum and a LFMA signal; additionally a DDPH pattern of paramagnetic nature is also included.

We will only center ourselves in EMR absorption, associated with the ferromagnetic resonance (FMR). This absorption satisfies the Larmor condition; when is applied to the case of a thin sheet with both negligible anisotropy field and the demagnetizing fields (Yildiz et al., 2002),

$$w = \gamma[(4\pi M + H_{efec}) \cdot H_{efec}]^{1/2} \qquad (1)$$

where ω is the microwave angular frequency (with ω = 2πf and f = 9.4 GHz), γ is the gyromagnetic ratio, H_{efec} is the effective magnetic field and M is the magnetization. The

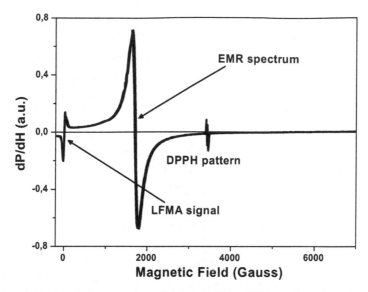

Figure 10. The derivative of microwave absorption from -1000 G to 8000 G, and which consists of an EMR spectrum and a LFMA signal (adapted from Montiel et al., 2005).

resonant condition implies that M–Ms with $H_{efec}= H_{dc}+H_{int}$, where H_{int} is the internal field. The saturation magnetization of the surface of the sample can be calculated from the resonance conditions as $4\pi M_s$ = 4741 G; which is close to the bulk saturation magnetization $4\pi M_s$=5250 G, the difference can be attributed to the fact that FMR is probing only the surface of the sample. Additionally, this absorption shows no hysteresis between the forward and backward field sweeps.

In Fig. 11, the temperature dependence of the EMR spectra can be observed. As temperature increases ΔH_{pp} becomes wider, due to new magnetic processes; where the dominant process is the dipole-dipole interaction associated with the paramagnetic phase, while the exchange interaction of the ferromagnetic order disappears when increasing the temperature. The dipole-dipole interaction has the effect of increasing the linewidth, while exchange interaction tends to narrow the absorption line. ΔH_{pp} as temperature function is shown in Fig. 12(a), the Curie temperature (T_c=482 K) is associated with the inflection point; where the derivative of the linewidth exhibited a maximum at T_c. A second EMR spectrum (SES) is detected after the magnetic transition, and it is associated with a second magnetic phase with a different Curie temperature; where this absorption mode is due to a nanocrystalline phase. The conductive behavior decreases due to the temperature increase, consequently, the absorption centers diminish and the EMR lineshape starts to become symmetrical. Also, the gain is one order of magnitude greater than at room temperature, and it is indicative of a reduced number of absorbing centers in the paramagnetic phase due to the high entropy of the system. Additionally, the temperature dependence of the resonant field is plotted in Fig. 12(b), where a shift in the resonant field is clearly observed. At 300 K, in ferromagnetic order, H_{res} corresponds to $H_{efec}= H_{dc}+H_{int}$, and as temperature increase H_{int} diminishes until $H_{int}=0$ at $T_c=482$ K in paramagnetic regime, with $H_{efec}= H_{dc}$.

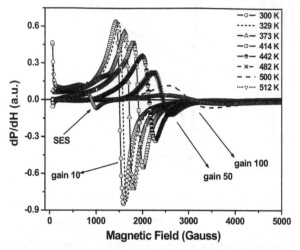

Figure 11. EMR spectra at different temperatures in amorphous ribbon $Co_{66}Fe_4B_{12}Si_{13}Nb_4Cu$; where the Curie temperature is detected at 482 K, and a decrease in microwave absorption is observed after magnetic transition.

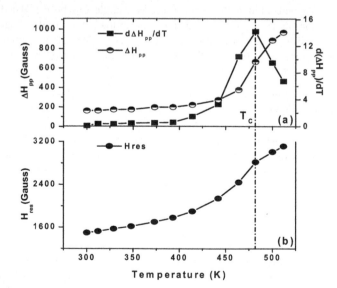

Figure 12. Temperature dependence of (a) ΔH_{pp} and (b) H_{res} in the amorphous ribbon $Co_{66}Fe_4B_{12}Si_{13}Nb_4Cu$ for EMR spectrum; solid lines are guides for the eye only. Also, Fig. 12(a) shows the derivative of ΔH_{pp} with the temperature, showing a maximum to Curie temperature.

3.1.2. LFMA technique

LFMA signal in amorphous ribbon is shown with more detail in Fig. 13(a). This signal is centered at zero magnetic field and shows an opposite phase to the EMR spectrum. The opposite phase is undoubtedly indicating that the microwave absorption has a minimum value at zero magnetic field, in contrast to the maximum value for EMR spectrum. LFMA signal has been interpreted as due to low-field spin magnetization processes (Beach & Berkowitz, 1994; Domínguez et al., 2002). We have correlated the LFMA signal with magnetoimpedance (MI) phenomena (Montiel et al., 2005), where Fig. 13(b) shows the MI response at 50 MHz for amorphous ribbon. The double peak clearly indicates low-field surface magnetization processes (Beach & Berkowitz, 1994) originated by the change in transversal permeability. The peak-to-peak width in MI is associated with the anisotropy field (H_K). In addition, Fig. 13(c) shows magnetometry measurements. The hysteresis loop is characterized by axial anisotropy, and a correlation between both experiments is observed on the basis of H_K. We compare measurements of LFMA, MI and magnetometry, in Fig. 13. A significant decrease of the microwave absorption (from H=16 G down to zero) is observed in LFMA measurements, whereas at the same fields, the magnetoimpedance measurement show that MI response is approaching saturation at field lower than 20 G. As the field decreases, a maximum is reached by MI, which corresponds to the anisotropy field (H=15.6 G). A further decrease of impedance is observed at zero field. As it is well known, MI is due to changes in the skin depth as a consequence of changes in the transversal permeability under the influence of the external H_{dc}. The change in domain structure, and therefore in spin dynamics, is also produced by H_{dc}, in a direct interaction with the axial anisotropy of the material. Experimentally, the maxima in MI signal coincide with the minimum and maximum of the LFMA signal and it can be associated with a common origin for both phenomena, where the magnetic processes in both phenomena are dependent of H_K.

The hysteresis effect of the LFMA signal appears to be due to a non-uniform surface magnetization processes. A ferromagnetic conducting system can absorb electromagnetic radiation energy and the efficiency of this absorption depends on the particular conditions such as: the magnetic domain structure, magnetic anisotropy, the orientation of the incident propagation vector radiation, its conductivity, its frequency and amplitude. This absorption can easily be modified by H_{dc}, which changes the magnetic susceptibility, the penetration depth, the magnetization vector, the domain structure and spin dynamics. Such changes can show hysteresis, as normally occurs in a domain structure subjected to dc fields lower than the saturating field. By cycling the H_{dc}, different irreversible domain configurations occur, and therefore a hysteresis effect can be obtained. These results clearly suggest that MI effect and LFMA signal represent the same responses to an external H_{dc}. MI and LFMA are due to domain structure and spin dynamics, and they can be understood as the absorption of electromagnetic energy by spin systems that are modified by domain configuration and strongly depends on H_K.

Let us consider an electromagnetic wave with both electric (**E**) and magnetic (**H**) fields. The time-average density of the power absorption (P), for a ferromagnetic conductor at high frequencies, can be expressed by the complex Poynting vector as: $P=\frac{1}{2}\text{Re}[E \times H^*]$ or

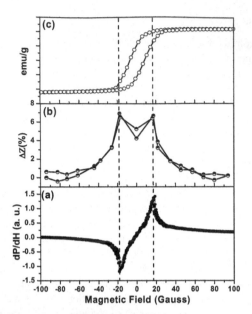

Figure 13. (a) LFMA signal, (b) MI signal at frequency of 50 MHz, and (c) VSM hysteresis loop for the amorphous ribbon Co$_{66}$Fe$_4$B$_{12}$Si$_{13}$Nb$_4$Cu (adapted from Montiel et al., 2005).

P=½Re[E·H*], where **H*** is the complex conjugate of **H**, and Re[x] the real part of the operator. Additionally, the ac surface impedance for a ferromagnetic conductor material is defined as the ratio of the fields at the surface: Z=E$_s$/H$_s$. Then the time-average density of the microwave power absorption can be written as P=½H$_s^2$Re(Z). The ac magnetic field H$_s$, in a ferromagnetic conductor at high frequency, is generated by a uniform current j=σE$_s$ (with σ the electrical conductivity) induced by the ac electric field E$_s$; and therefore H$_s$ is constant to changes of an applied static magnetic field.

Therefore, we can establish a relation between the field derivative (dP/dH) of the microwave power absorption and the rate of change of Re (Z) with an applied static magnetic field, H:

$$dP/dH = \left(H_s^2/2\right)\left[dRe(Z)/dH\right] \tag{2}$$

For a good magnetic conductor Z= (1+j)/σδ, with the classical skin depth 1/δ= (ωμσ/2)^{½} and μ the permeability. The magnetoimpedance is defined as the change of the impedance of a magnetic conductor subjected to an ac excitation current, under the application of a static magnetic field H$_{dc}$; it is a very similar phenomenon to the one involved in the microwave power absorption. At high-frequencies (microwaves) and due to the skin depth effect, only the surface impedance is involved.

LFMA signal can be used to detect magnetic order and to determine Curie temperature, because the appearance of LFMA signal has been widely accepted as a signature of the onset of the ferromagnetic transition (Montiel et al., 2004; Alvarez et al., 2010; Gavi et al., 2012). Therefore, it is possible to establish that for temperatures above the Curie temperature, the long-range magnetic order is completely lost and LFMA signal disappears. LFMA signal shows a decrease in the intensity as temperature is approached to T_c, and finally disappeared at $T \geq T_c$. Fig. 14 shows the temperature dependence of the LFMA signal, it is necessary to mention that LFMA signal is located around zero field for all temperatures. At room temperature, the LFMA signal has a phase opposite to EMR spectrum. As temperature increases, for $T \geq 373$ K, LFMA signal invests its phase until disappearing. This behavior is correlated with the long-range order in the ferromagnetic state and with the temperature dependence of the anisotropy field. The phase change has been observed previously in nickel around Curie transition, Nabereznykh & Tsindlekht (1982), and it can be explained by means of magnetic fluctuations; and they are associated with the electric properties of the material. LFMA signal showed a decrease in ΔH_{LFMA} and hysteresis remains until T_c is reached.

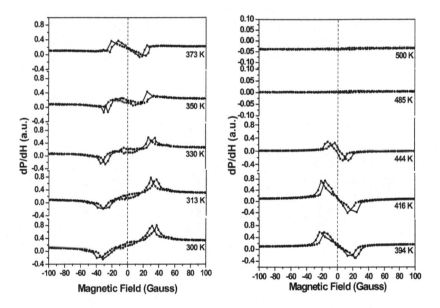

Figure 14. LFMA signals at different temperatures in the amorphous ribbon $Co_{66}Fe_4B_{12}Si_{13}Nb_4Cu$. At 485 K the LFMA signal disappears, indicating the transition from the ferromagnetic order to paramagnetic phase.

3.2. Curie and Yafet-Kittel transitions (from ferrimagnetic order to a Yafet-Kittel-type ordering)

The polycrystalline Ni-Zn ferrites ($Ni_{1-x}Zn_xFe_2O_4$, $0 \leq x \leq 1$) are an important family of solid solutions with a remarkable variety of magnetic properties and applications

(Ravindaranathan et al., 1987). This solid solution crystallizes in a cubic spinel-type structure, see Fig. 15(a), where Zn ions normally are located in tetrahedral sites (A-sites) and Ni ions have a marked preference to occupy the octahedral sites (B-sites), while Fe ions are distributed among both sites types. The antiferromagnetic superexchange interaction (A-O-B) is the main cause of the cooperative behavior of the magnetic moments in Ni-Zn ferrites below their Curie temperature. In a great variety of experimental observations (Satya Murthy et al., 1969; Pong et al., 1997; Akther Hossain et al., 2004) have found that for x≤0.5 the resultant of the magnetic moments in the A and B sites has a classic collinear arrangement, see Fig. 15(b); while that for x>0.5 a non-collinear arrangement of the magnetic moments is employed to explain these behaviors. The previous behaviors are because the superexchange interaction B-O-B begins to be comparable with A-O-B interaction, and the arrangement of the magnetic moments shows a Yafet-Kittel-type canting (Yafet & Kittel et al., 1952). Also, the transition temperature from a ferrimagnetic ordering (collinear arrangement) to a Yafet-Kittel-type magnetic ordering (non-collinear arrangement) is called Yafet-Kittel temperature (T_{YK}); in particular, $Ni_{0.35}Zn_{0.65}Fe_2O_4$ ferrite has a T_{YK} smaller than the Curie temperature (Satya Murthy et al., 1969; Akther Hossain et al., 2004). The polycrystalline $Ni_{0.35}Zn_{0.65}Fe_2O_4$ ferrite was prepared by two different methods: the conventional classical ceramic method known as the solid-state reaction, and the co-precipitation method.

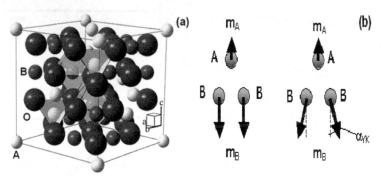

Figure 15. (a) Schematic representation of unit cell structure for Ni-Zn ferrites; where A and B are tetrahedral and octahedral sites, respectively. (b) Spin orientation on the A and B sites, for the collinear (to left) and the non-collinear (to right) model, in Ni-Zn ferrites; where α_{YK} is the Yafet-Kittel angle (adapted from Alvarez et al., 2010).

3.2.1. EMR technique

Fig. 16 shows the EMR spectra of the Ni-Zn ferrite prepared by solid-state reaction at different temperatures. For all temperature range, EMR spectra exhibit a broad signal, but their lineshape change with a shift in H_{res} when varying the temperature. Beginning to low temperature, an asymmetric mode (FMR signal) is observed and it gradually changes to a symmetric mode (EPR signal) when increasing the temperature. This change is associated with the transition from a ferrimagnetic order to a paramagnetic phase; i.e. the evolution from a FMR spectrum to an EPR spectrum is used to determine the Curie temperature (T_c) in Ni-Zn ferrites (Montiel et al., 2004; Wu et al., 2006; Priyadharsini et al., 2009; Alvarez et al., 2010). Additionally, EMR spectra exhibit an additional absorption at low magnetic field, this new absorption mode is a LFMA signal which will be discussed with more detail in the following section.

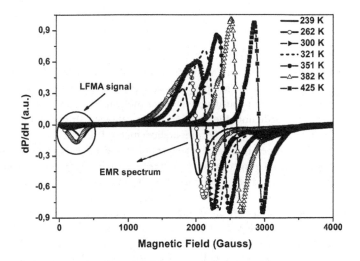

Figure 16. EMR spectra of the Ni-Zn ferrite prepared by solid-state reaction, for selected temperatures in the 239-425 K temperature range; circle shows LFMA signal.

In a polycrystalline magnetic material, the resonance condition for FMR signal is expressed as:

$$w = \gamma H_{res} \tag{3}$$

with $H_{res}=H_{dc}+H_{int}$, and H_{int} is the internal field which is the combination of several factors associated with the long-range order in the ferrite (Schlomann, 1958): the anisotropy field (H_K), the porosity field (H_p), the field due to eddy currents (H_e) and the demagnetization field (H_d). Additionally, the inhomogeneities in Ni-Zn ferrites also can contribute to internal field, and they are associated with differences in sites occupancy by cations. Other source of inhomogeneity in the internal field is the disorder in the site occupancy. Even if the occupancy of sites is well determined (i.e., in Ni-Zn ferrite, all Zn cations on A sites, all Ni cations on B sites), there can be an inhomogeneous distribution of each of them on the sites. EMR spectra can be slightly different when this occupancy of sites is not strictly homogeneous, since some terms of H_{int} are not exactly the same for all the absorbing centers. It is possible to change the cations distribution in the ferrites by means of thermal treatments or when preparing samples of the same composition but with different synthesis methods.

H_{res} as a function of temperature for the conventional and co-precipitate Ni-Zn ferrites are shown in Fig. 17(a). The values for conventional ferrite are lightly higher than those of the co-precipitate ferrite. This difference can be due to inhomogeneities associated with a different distribution of cations, and that it is originated by synthesis method. For both samples, the increment of H_{res} as temperature increases is due to decrement of the internal field, i.e. in the ferrimagnetic order H_{int} is added to the applied field and the resonance condition is reached at low values of H_{dc}. In contrast, in the paramagnetic phase, the necessary magnetic field to satisfy the resonance condition has to be supplied entirely by the applied field, $H_{int}= 0$ and $H_{res}= H_{dc}$; i.e. when increasing the temperature the progressive disappearance of H_{int} is associate with the lost long-range order.

ΔH_{pp} can be due to several factors (Srivastava & Patni, 1974), in particular for polycrystalline samples the linewidth is due to: the sample porosity (ΔH_{por}), the magnetic anisotropy (ΔH_K), the eddy currents (ΔH_{eddy}), and the magnetic demagnetization (ΔH_{des}). It is necessary to mention that there is a broadening due to variations in cations distribution on the A and B sites, and it is highly dependent of the preparation method. In Fig. 17(b), we show the behavior of ΔH_{pp} with temperature for the conventional and co-precipitate Ni-Zn ferrites. For all the temperatures, ΔH_{pp} in co-precipitate ferrite is higher than for the conventional ferrite. This behavior can be due to differences in the microscopic magnetic interactions inside the samples, mainly the interparticle magnetic dipole interaction and the superexchange interaction; and they are originated by different cations distributions in the samples, due to synthesis method. The magneto-crystalline anisotropy has a strong contribution to ΔH_{pp} and we can have the following approximation $\Delta H_{pp}=\Delta H_K=K_1/2M_s$, i.e. for a system of randomly oriented crystallites, the contribution of the anisotropy field is

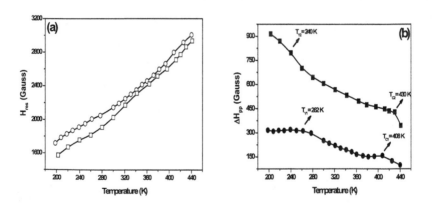

Figure 17. (a) Temperature dependence of H_{res} in the 200-440 K temperature range for the ($\circ$) conventional and ($\square$) co-precipitate Ni-Zn ferrites. (b) Temperature dependence of ΔH_{pp} in the same temperature range for the ($\bullet$) conventional and ($\blacksquare$) co-precipitate Ni-Zn ferrites. The curves connecting points are only guides for the eye.

dominant. For both ferrites, the decrease in ΔH_{pp} as temperature increases is associated to a weakening of the magneto-crystalline anisotropy as T approaches T_c (Byun et al., 2000). The magnetic transition (ferri-paramagnetic) appears as an inflection point in plot ΔH_{pp} vs. temperature, as is shown in Fig. 17(b). The inflection points at $T_{c1} \sim 408$ K (conventional) and $T_{c2} \sim 430$ K (co-precipitate) are associated with Curie temperature of each sample. We observed that the Curie transition is higher in the co-precipitate Ni-Zn ferrite. As T_c is an intrinsic property, that it depends entirely on the ferrite composition, the difference in Curie temperature between both ferrites also suggests a different occupancy between the A and B sites. To high temperature, the long-range magnetic order is completely lost except for some short range order islands in the material that contribute strongly in the broadening of the EMR spectrum. On the other hand, in Fig. 17(b), a second inflection point is clearly observed at $T_{Y1} = 262$ K and $T_{Y2} = 240$ K, in conventional and co-precipitate ferrites respectively. This behavior is attributed to a non-collinear arrangement of the magnetic moments in the A and B sites, i.e. it is due to a Yafet-Kittel-type ordering of the magnetic moments in both samples; the difference of temperatures ($T_{Y1} > T_{Y2}$) between both ferrites is indicative of an inhomogeneous distribution on occupancy of the B sites.

3.2.2. LFMA technique

In Fig. 18(a), we show LFMA spectra for conventional ferrite in the 208-408 K temperature range. This microwave absorption, around zero field, is far from the resonance condition given by eq.(3), i.e. the sample is in an unsaturated state; therefore, this absorption is associated with interaction between the microwave field and the dynamics of the magnetic domains structure in the sample. For $T > T_{c1}$ (= 408 K), LFMA spectra exhibit a linear behavior with a positive slope and non-hysteretic traces, and which is associate with paramagnetic phase, i.e. the long-range order has completely disappeared in the sample. For T_{YK1}(= 261 K) $\leq T \leq T_{c1}$, LFMA spectra show an antisymmetrical shape around zero field, displaying a clear hysteresis upon cycling the field, see Fig. 18(a); and they have the same phase of the EMR spectra, indicating that this absorption has a maximum value at zero field. For this temperature range, it can be observed that ΔH_{LFMA} increases when the temperature decreases, as can be seen in Fig. 18(b), for conventional and co-precipitate ferrites; this behavior indicates an increase in the ferromagnetically coupled superexchange interactions in the samples, suggesting that ΔH_{LFMA} is determined by the magnetic anisotropy and by the demagnetizing field (Montiel et al., 2004, 2005; Alvarez et al., 2008, 2010). But between both samples, the widths difference can be due to different distributions of cations in B sites, originating changes in the antiferromagnetic superexchange interactions; being more intense in the conventional Ni-Zn ferrite.

For $T < T_{YK1}$, an additional absorption mode also centered at zero field is observed in Fig. 18(a), and that it is suggested by high distortion of the LFMA signal at low temperature; this new absorption mode is more evident in co-precipitate ferrite (Alvarez et al., 2010). This signal exhibits an opposite phase (out-of-phase) with regard to EMR spectra; indicating that this microwave absorption has a minimum value at zero field. The presence of an out-of-phase signal has been correlated with the occurrence of a ferromagnetic order. It can be assumed that a ferromagnetic arrangement is related with this signal, while a ferrimagnetic structure leads to the opposite result. Therefore, this out-of-phase signal can be associated with the appearance of a ferromagnetic arrangement of the magnetic moments in the sample; where a Yafet-Kittel-type canting of the magnetic moments in the B sites can provide this ferromagnetic component. Additionally, in this temperature region, ΔH_{LFMA} increases continuously with temperature decrease, but now with a higher change rate; this quick broaden is due to a build-up of the short-range magnetic correlations preceding to magnetic transition. For the above-mentioned, we propose that the phase transition at low temperature is due to a Yafet-Kittel-type magnetic ordering of the moments in the B sites, with onset at T_{YK1}; i.e. for $T<T_{YK1}$, the parallel arrangement of the B sites is modified, and a non-collinear arrangement of the magnetic moments in the A and B sites appears, leading to a change in the microwave absorption regime. It is necessary to mention that a similar behavior is observed in co-precipitate ferrite but with $T<T_{YK2}$ (= 239 K), see Fig. 18(b). The relevant temperatures are different but the dynamics of the magnetic moments is similar, detecting the Yafet-Kittel-type magnetic ordering around T_{YK2}, in a good correspondence with EMR technique. The difference of temperatures ($T_{YK1}>T_{YK2}$) between both ferrite is also indicative of a different occupancy of cations in the A and B sites.

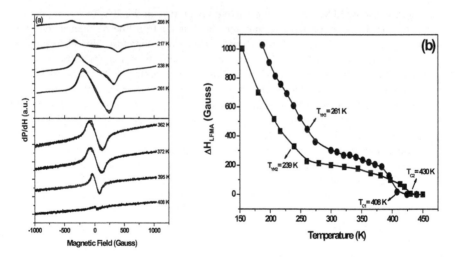

Figure 18. (a) LFMA spectra of the conventional Ni-Zn ferrite in the 362-408 K (down) and 208-261 K (up) temperature ranges. (b) Behavior of ΔH_{LFMA} for the (●) conventional and (■) co-precipitate Ni-Zn ferrites, as a function of temperature in the 150-450 K temperature range.

3.2.3. MAMMAS technique

We used MAMMAS technique to detect the ferri-paramagnetic transition in both ferrites sample. MAMMAS responses are shown in Fig. 19, where the measurements are carried out heating and cooling the samples; with the purpose of looking a change associated with ferri-paramagnetic transition. For the sample of solid-state reaction, during heating (cooling), this signal increases monotonically as temperature increases (decreases) from 320 K (440 K), reaching a maximum value at T_{p1}=379 K (T^*_{p1}=355 K). As temperature increases (decreases) further, $T>T_{p1}$ ($T<T^*_{p1}$), the MAMMAS response decreases and another magnetic process sets-in, modifying its microwave absorption and this suggests a magnetic transition. An observed interesting feature in MAMMAS response is the absence of thermal hysteresis in the heating and cooling cycles, and it is only observed when $T\leq T_{H1}$(= 396 K); this merging point (T_{H1}) indicates the onset of the magnetic ordering. The T_{H1} value is in a very good agreement with the value of Curie temperature detected by the EMR and LFMA measurements. A similar MAMMAS response has been observed in co-precipitate ferrite in the 350-440 K temperature range, see the inset of the Fig. 19. The relevant temperatures are different but the dynamics of the magnetic moments is similar, detecting the ferri-paramagnetic transition around T_{H2}=424 K, in a good correspondence with EMR and LFMA techniques.

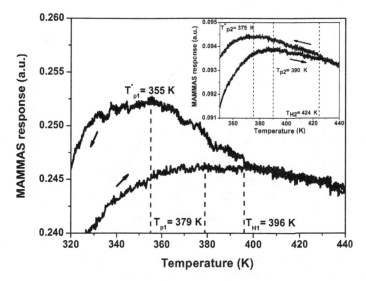

Figure 19. MAMMAS response of the conventional Ni-Zn ferrite in the 320-440 K temperature range; the inset shows the MAMMAS response of the co-precipitate Ni-Zn ferrite in the 350-440 K temperature range.

MAMMAS responses for these two ferrites, in 150-300 K temperature range, give us additional information on the magnetic transition at low temperature (see Fig. 20). Beginning to 300 K, these responses exhibit a continuous decrease to a minimum value at T_{m1}= 260 K and T_{m2}= 240 K, in conventional and co-precipitate ferrites respectively; and these absorptions increases when continuing diminishing the temperature. This feature points to a change in the microwave absorption regime due to a change in the magnetic structure

(Alvarez et al., 2010); revealing the appearance of a new population of absorbing centers, and which it also is suggested from EMR and LFMA measurements. All this profile shows that this change appears progressively as temperature is diminished, i.e. it is not a sharp change as could be expected from a structural phase transition. We associate this behavior with the transition from the collinear magnetic structure, $T>T_{m1}$ and $T>T_{m2}$, to the non-collinear (Yafet-Kittel-type) structure. Therefore, the MAMMAS responses depend on the thermal dependence of the magnetic moments dynamics, and the intensity of these signals follow the variations on the number of absorption centers (as is suggested by the EMR parameters), in turn is controlled by the establishment of the Yafet-Kittel-type canting of magnetic moments in the B sites at low temperature.

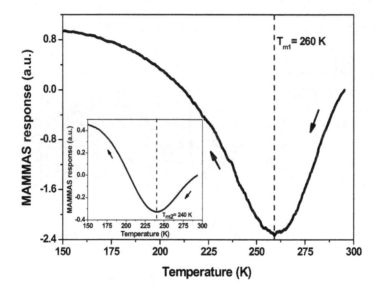

Figure 20. MAMMAS response of the conventional Ni-Zn ferrite in the 150-300 K temperature range; the inset shows the MAMMAS response of the co-precipitate Ni-Zn ferrite for same temperature range.

3.3. Néel transition (from a paramagnetic phase to antiferromagnetic ordering)

Lead iron tungstate, $Pb(Fe_{2/3}W_{1/3})O_3$ (PFW), shows an Néel transition around 350-380 K (Smolenskii et al., 1964; Feng et al., 2002; Ivanova et al., 2004); this magnetic ordering is due to superexchange interaction between the Fe ions through the O ions. For this study, PFW powders has been prepared through columbite precursor method (Zhou et al., 2000), where the purity of the powders was checked by means of XRD; all observed reflection lines are indexed as a cubic perovskite-type structure, in a good agreement with the standard data for PFW powders.

3.3.1. EMR technique

Fig. 21 shows the EMR spectra recorded in the 294-423 K temperature range. For 355 K <T≤ 423 K, we observed a single broad Lorentzian line due to spin of the Fe^{+3} ions. In this lineshape kind, the derivative of the microwave power absorption with respect to the static field (dP/dH) can be fitted into two-component Lorentzian, accounting for the contributions from the clockwise and anticlockwise rotating components of the microwave magnetic field (Joshi et al., 2002). For T ≤ 355 K, when the temperature goes diminishing, the absorption mode changes toward a broad asymmetric line of Dyson-type (Dyson, 1955; Feher & Kip, 1955) and their intensity diminishes; this Dyson lineshape is associated with a conductive contribution. This lineshape is a combination of an absorption component and other dispersion component of a symmetric Lorentzian mode, originating an additional parameter: the A/B ratio, i.e. the ratio of the amplitude of the left peak to that of the right peak of the EMR spectrum. Thus, EMR spectra are fitted to a functional form similar to the one used by Ivanshin et al. (2000). Additionally, for this temperature range, a second EMR mode (signal 2) is also observed; being more evident at room temperature, see Fig. 21.

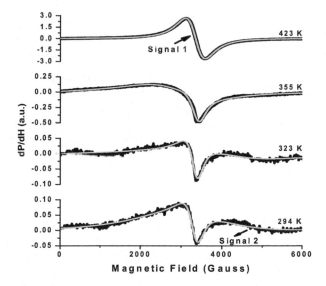

Figure 21. EMR spectra of PFW powders for selected temperatures; the solid lines correspond to the fits obtained from a functional form similar to the one used by Ivanshin et al. (2000).

In Fig. 22, the temperature dependences of the EPR parameters for signal 1 are plotted. ΔH_{PP} as a temperature function is shown in Fig. 22(a) for PFW powders. Starting from 423 K, as temperature decreases, ΔH_{PP} decreases continuously until 323 K; exhibiting a minimum at this temperature. This narrowing in ΔH_{PP} indicates an increase in the superexchange interactions in the sample, because the superexchange mechanism tend to narrow the absorption line; and it can be due to magnetic fluctuations, i.e. fluctuations in the establishment of the long-range order that precedes the transition to antiferromagnetic order. When continuing diminishing the temperature, T<323 K, ΔH_{PP} shows a weak increase until 294 K, this increase is indicative of a weak ferromagnetic behavior in the PFW; similar behaviors are observed in others magnetoelectric materials (Alvarez et al., 2006, 2010, 2012).

Fig. 22(b) shows the behavior of the g-factor vs. temperature for the signal 1, which is estimated from H_{res}, with $g = h\nu/\mu_B H_{res}$; where h is the Planck constant, ν is the frequency and μ_B is the Bohr magneton. EMR spectra give g-factor smaller than for a free electron (= 2.0023), along the entire temperature range. This behavior can be explained through the spin value of the Fe^{3+} ions (S=5/2) and to changes in the spin-orbit coupling; where the effective g-factor (g_{eff}) of a paramagnetic center is given by $g_{eff} = g(1 \pm \kappa/\Delta)$, where Δ is the crystal-field splitting and κ is the spin-orbit coupling constant. The g-factor shows a weak decrease in the 423-365 K temperature range and then the g-factor continues diminishing, but this time with a higher change rate, reaching a minimum value at 344 K ($g_{min} = 1.8491$). This fast decrease can be due to build-up of magnetic correlations preceding the transition to the long-range antiferromagnetic ordering at Néel temperature ($T_N \sim 344$ K). For T<T_N, the g-factor increases until 294 K and this behavior is an indication of a weak ferromagnetism in the PFW; this behavior can be due to a canting of the sublattices of Fe^{3+} ions in the antiferromagnetic matrix, generating an effective magnetic moment.

The temperature dependence of the A/B ratio is shown in Fig. 22(c). From this plot can be seen that, starting from 423 K to close to 378 K, the A/B ratio remains essentially constant at a value of 1. This value indicates that the paramagnetic centers are static and also suggests a strong dipolar interaction between Fe^{3+} ions. Further, the A/B ratio continually diminishes until a minimum value to 344 K; indicating a dispersion contribution and it suggests a conduction effect in the sample. As the temperature is decreased further, 294 K≤T<344 K, the A/B ratio increases toward a near value of 1, due to a decrease of the conductivity in sample. Recently, electric mensurations were carried out on PFW samples (Eiras et al. 2010; Fraygola et al. 2011), and which indicate a conductive contribution associated with an electronic-hoping mechanism, in a good correspondence with Dyson lineshape of the EMR spectra.

A second absorption mode (signal 2) is clearly observed at room temperature and it is associated with the presence of a fraction of Fe^{2+} ions, where oxygen atoms deficiency generate a state of mixed valency; originating a strong magnetic dipolar interaction between Fe^{2+} and Fe^{3+} ions, and that produces a broaden in absorption mode. In inset of the Fig. 22(a), we show the ΔH_{PP} behavior as a temperature function for the signal 2. For T<378 K, when diminishing the temperature, the broadening of signal 2 is indicative of an increase of the interactions with Fe^{2+} ions. Additionally, a change in slope to 334 K is also observed; see the inset of the Fig. 22(a), where this feature can be associated with the para-antiferromagnetic transition.

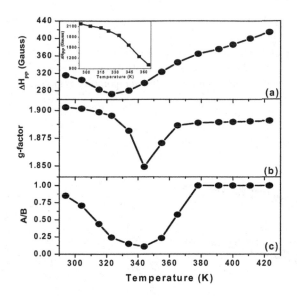

Figure 22. Temperature dependence of (a) ΔH_{PP}, (b) the g-factor and (c) the A/B ration of PFW powders for the signal 1; the inset of the Fig. 22(a) shows the temperature dependence of ΔH_{PP} for the signal 2. Solid lines are guides for the eye only.

3.3.2. LFMA technique

In Fig. 23(a), we show the LFMA signal in the 294-423 K temperature range. For all temperatures, LFMA signal exhibits two antisymmetric peaks about zero magnetic field, with opposed phase to EMR spectrum, and a clear hysteresis of this signal appears on cycling the field. This strongly contrasts with the LFMA signal observed in others magnetoelectric materials (Alvarez et al., 2007, 2010, 2012), where these absorptions are lineal and non-hysteretic. The hysteresis feature has been associated with low field magnetization processes in ferromagnetic materials (Montiel et al., 2004; Alvarez et al., 2010; Gavi et al., 2012), suggesting the presence of a magnetic component in the material (Fraygola et al., 2011). With the help of a reference line, clearly one can observe that the LFMA signal has a lineal absorption component, see Fig. 23(a); where their slope is a temperature function. Fig. 23(b) shows the slope behavior of the lineal component for the 294-423 K temperature range. The slope increases monotonically when the temperature decreases from 423 K, reaching a maximum value at 365 K; this increasing behavior is characteristic of a paramagnetic phase. As the temperature is decreased further, T<365 K, the slope decreases very fast with decreasing temperature until T_{min}= 334 K; in this region the quantity of absorbing centers diminishes considerably due to the process of antiparallel spin alignment. Below T_{min}, the slope has an approximately lineal increase, where this behavior is a signature of the weak ferromagnetism in this temperature region (Alvarez et al., 2007, 2010, 2012).

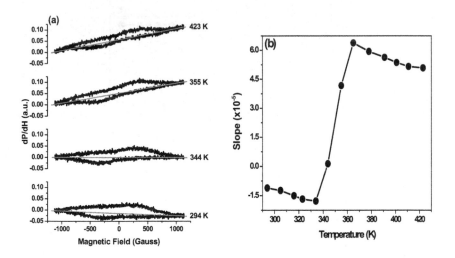

Figure 23. (a) LFMA signal for selected temperatures of PFW powders, where the straight lines are only a help to visualize the lineal component of the LFMA signal. (b) The slope temperature dependence of the lineal component for LFMA signal in the 294-423 K temperature range; the solid lines are guides for the eye only.

4. Conclusions

In this work is shown that EMR is the most powerful spectroscopic method available to determine the magnetic transitions in the materials. LFMA and MAMMAS techniques provide information on the dependences in temperature and magnetic field of the non-resonant microwave absorption. More important, these techniques can distinguish between different dissipative dynamics of microwave absorbing centers, providing valuable information about the nature of magnetic ordering within materials. We have shown that all these techniques are powerful tools for the research of magnetic materials at microwave frequencies.

Author details

H. Montiel
Centro de Ciencias Aplicadas y Desarrollo Tecnológico, Universidad Nacional Autónoma de México, Del. Coyoacán, México DF 04510, México

G. Alvarez
Escuela Superior de Física y Matemáticas del IPN, U.P.A.L.M, Edificio 9, San Pedro Zacatenco, México DF 07738, México

Acknowledgement

G. Alvarez acknowledges research support in the laboratory of magnetic mensurations and biophysics of ESFM-IPN-Mexico. The authors would like to thank R. Zamorano by the use of the EMR spectrometer. Support from project PAPIIT-UNAM No. IN111111 is gratefully acknowledged.

5. References

Akther Hossain A.K.M., Seki M., Kawai T., Tabata H., (2004) J. Appl. Phys. 96, 1273.

Alvarez G., Zamorano R., (2004) J. Alloys Compd. 369, 231.

Alvarez G., Font R., Portelles J., Valenzuela R., Zamorano R., (2006) Physica B 384, 322.

Alvarez G., Font R., Portelles J., Zamorano R., Valenzuela R., (2007) J. Phys. Chem. Solids 68, 1436.

Alvarez G., Montiel H., Cos D., García-Arribas A., Zamorano R., Barandiarán J.M., Valenzuela R., (2008) J. Non-Cryst. Solids 354, 5195.

Alvarez G., Font R., Portelles J., Raymond O., Zamorano R., (2009) Solid State Sci. 11, 881.

Alvarez G., Montiel H., Barron J.F., Gutierrez M.P., Zamorano R., (2010) J. Magn. Magn. Mater. 322, 348.

Alvarez, G., Cruz M.P., Durán A.C., Montiel H., Zamorano R., (2010) Solid State Commun. 150, 1597.

Alvarez G., Peña J.A., Castellanos M.A., Montiel H., Zamorano R., (2012) Rev. Mex. Fis. S 58(2), 24.

Andrzejewski B., Kowalczyk A., Stankowski J., Szlaferek A., (2004) J. Phys. Chem. Solids 65, 623.

Beach R.S., Berkowitz A. E., (1994) J. Appl. Phys. 76, 6209.

Bele P., Brunner H., Schweitzer D., J. Keller H., (1994) Solid State Commun. 92, 189.

Bhat S.V., Ganguly P., Ramakrishnan T.V., Rao C.N.R., (1987) J. Phys. C: Solid State Phys. 20, L559.

Bhide M.K., Kadam R.M., Sastry M.D., Ajay Singh, Shashwati Sen, Aswal D.K., Gupta S.K., Sahni V.C., (2001) Supercond. Sci. Technol. 14, 572.

Blazey K.W., Müller K.A., Bednorz J.G., Berlinger W., Amoretti G., Buluggiu E., Vera A., Matacotta F.C., (1987) Phys. Rev. B 36, 7241.

Bohandy J., Suter J., Kim B.F., Moorjani K., Adrian F.J., (1987) Appl. Phys. Lett. 51, 2161.

Byun T.Y., Byeon S.C., Hong K.S., Kyung C., (2000) J. Appl. Phys. 87, 6220.

Domínguez M., García-Beneytez J.M., Vázquez M., Lofland S.E., Baghat S.M., (2002) J. Magn. Magn. Mater. 249, 117.

Dyson F.J., (1955) Phys. Rev. 98, 349.

Eiras J.A., Fraygola B.M., Garcia D., (2010) Key Engineering Materials 434-435, 307.

Feher G., Kip A.F., (1955) Phys. Rev. 98, 337.

Feng L., Ye Z.G., (2002) J. Solid State Chem. 163, 484.

Fraygola B.M., Coelho A.A., Garcia D., Eiras J.A., (2011) Materials Research 14, 434.

Gavi H., Ngomb B.D., Beye A.C., Strydom A.M., Srinivasu V.V., Chaker M., Manyala N., (2012) J. Magn. Magn. Mater. 324, 1172.

Healy D.W., (1952) Phys. Rev. 86, 1009.

Hirotake Kajii, Hisashi Araki, Zakhidov A.A., Kazuya Tada, Kyuya Yakushi, Katsumi Yoshino, (1997) Physica C 277, 277.

Ivanova S.A., Eriksson S.G., Tellgrend R., Rundlöf H., (2004) Mater. Res. Bull. 39, 2317.

Ivanshin V.A., Deisenhofer J., Krug von Nidda H.-A., Loidl A., Mukhin A., Balbashov J., Eremin M.V., (2000) Phys. Rev. B 61, 6213.

Joshi J.P., Gupta R., Sood A.K., Bhat S.V., Raju A.R., Rao C.N.R., (2002) Phys. Rev. B 65, 024410.

Khachaturyan K., Weber E.R., Tejedor P., Stacy A.M., Portis A.M., (1987) Phys. Rev. B 36, 8309.

Kheifets A.S., Veinger A.I., (1990) Physica C 165, 491.

Kim B.F., Moorjani K., Adrian F.J., Bohandy J., (1993) Materials Science Forum 137-139, 133.

Montiel H., Alvarez G., Gutierrez M.P., Zamorano R., Valenzuela R., (2004) J. Alloys Compd. 369, 141.

Montiel H., Alvarez G., Betancourt I., Zamorano R., Valenzuela R., (2005) Appl. Phys. Lett. 86, 072503.

Montiel H., Alvarez G., Betancourt I., Zamorano R., Valenzuela R., (2006) Physica B 384, 297.

Montiel H., Alvarez G., Zamorano R., Valenzuela R., (2007) J. Non-Cryst. Solids 353, 908.

Montiel H., Alvarez G., Zamorano R., Valenzuela R., (2008) J. Non-Cryst. Solids 354, 5192.

Moorjani K., Bohandy J., Adrian F.J., Kim B.F., (1987) Phys. Rev. B 36, 4036.

Nabereznykh V.P., Tsindlekht M.I., (1982) JETP Lett. 36, 157.

Niebling U., Steinl J., Schweitzer D., Strunz W., (1998) Solid State Commun. 106, 505.

Okamura T., Torizuka Y., Kojima Y., (1951) Phys. Rev. 82, 285.

Okamura T., (1951) Nature 168, 162.

Owens F.J., (1997) J. Phys. Chem. Solids 58, 1311.

Owens F.J., (2001) Physica C 363, 202.

Pacher N., Deisenhofer J., Krung von Nidda H.-A., Hemmida M., Jeevan H.S., Gegenwart P., Loidl A., (2010) Phys. Rev. B 82, 054525.

Padam G.K., Ekbote S.N., Tripathy M.R., Srivastava G.P., Das B.K., (1999) Physica C 315, 45.

Padam G.K., Arora N.K., Ekbote S.N., (2010) Mater. Chem. Phys. 123, 752.

Panarina N.Y., Talanov Y.I., Shaposhnikova T.S., Beysengulov N.R., Vavilona E., Behr G., Kondrat A., Hess C., Leps N., Wurmehl S., Klinger R., Kataev V., Büchner B., (2010) Phys. Rev. B 81, 224509.

Pong W.F., Chang Y.K., Su M.H., Tseng P.K., Lin H.J., Ho G.H., Tsang K.L., Chen C.T., (1997) Phys. Rev. B 55, 11409.

Priyadharsini P., Pradeep A., Sambasiva Rao P., Chandrasekaran G., (2009) Mater. Chem. Phys. 116, 207.

Ravindaranathan P., Patil K.C., (1987) J. Mater. Sci. 22, 3261.

Satya Murthy N.S., Natera M.G., Youssef S.I., Begum R.J., Srivastava C.M., (1969) Phys. Rev. 181, 969.

Schlomann E., (1958) J. Phys. Chem. Solids 6, 257.

Shaltiel D., Bezalel M., Revaz B., Walker E., Tamegai T., Ooi S., (2001) Physica C 349, 139.

Smolenskii G.A., Bokov V.A., (1964) J. Appl. Phys. 35, 915.

Srivastava C.M., Patni M.J., (1974) J. Magn. Res. 15, 359.

Stankowski J., Piekara-Sady L., Kempinski W., (2004) J. Phys. Chem. Solids 65, 321.

Topacli C., (1996) J. Supercond. 9, 263.

Topacli C., (1998) Physica C 301, 92.

Velter-Stefanescu M., Totovana A., Sandu V., (1998) J. Supercond. 11, 327.

Velter-Stefanescu M., Duliu O.G., Mihalache V., (2005) J. Optoelectron. Adv. M. 7, 1557.

Wu K.H., Shin Y.M., Yang C.C., Wang G.P., Horng D.N., (2006) Mater. Lett. 60, 2707.

Yafet Y., Kittel C., (1952) Phys. Rev. 87, 290.

Yildiz F., Rameev B.Z., Tarapov S.I., Tagirov L.R., Aktas B., (2002) J. Magn. Magn. Mater. 247, 222.

Zakhidov A.A., Ugawa A., Imaeda K., Yakushi K., Inokuchi H., (1991) Solid State Commun.
    79, 939.
Zhou L., Vilarinho P.M., Bptista J.L., Fortunato E., (2000) J. Eur. Ceram. Soc. 20, 1035.

Chapter 6

Microwave Absorption
in Nanostructured Spinel Ferrites

Gabriela Vázquez-Victorio, Ulises Acevedo-Salas and Raúl Valenzuela

Additional information is available at the end of the chapter

1. Introduction

Magnetic Nanoparticles

Magnetic nanoparticles (MNPs) are playing a crucial role in an extensive number of potential applications and science fields. Nanotechnology industry is rapidly growing with the promise that it will lead to significant economic and scientific impacts on a wide range of areas, such as health care, nanoelectronics, environmental remediation. MNPs are mostly ferrites, i.e, transition metal oxides with ferric ions as main constituent. Although the magnetic properties of ferrites [1] are less intense than metal's, especially saturation magnetization, ferrites possess a large chemical stability (corrosion resistance), high electrical resistivity, and extended applicability at high magnetic field frequencies.

The use of MNPs for biological and clinical applications [2] is undoubtedly one of the most challenging research areas in the field of nanomaterials, involving the organized collaboration of research teams formed by physicists, chemists, biologists, physiciens. The advantages of MNPs are based on their nanoscale size, large surface area, tailoring of magnetic properties and negligible side effects in living tissues. These applications include drug delivery [3], magnetic hyperthermia [4], magnetic resonance imaging [5], biosensors [6]. A field related with microwave absorption is electromagnetic interference EMI [7], as the number of electromagnetic radiation sources has growth at an exponential rate. MNPs have found applications also in environmental fields, such as soil remediation [8] and heavy metal removal [9], as MNPs provide high surface area and specific affinity for heavy metal adsorption from aqueous systems.

The reduction in scale leads to strong changes in macroscopic properties. The main reason can be attributed to the enhanced importance of the surface atom fraction as compared with core atom fraction, as the material becomes a nanoparticle. A simple estimate reveals that for a ~100 nm nanoparticle, surface atom fraction is about 6% of the total NP atoms, while for a ~5 nm NP this fraction can attain 78% [10]. Surface layer of materials exhibits different properties simply because these atoms have a very different structure than the core's. Surface atoms, for instance, have a reduced coordination number (unsatisfied bonding), crystal defects and modified crystal planes ("broken symmetry"). In the case of magnetically ordered materials (ferro, ferri, antiferromagnetic phases), additionally, several magnetic properties critically change at the nanometric scale. These properties are, for instance, the change from multidomain to single domain magnetic structure, domain wall thickness, the decrease in anisotropy energy giving rise to superparamagnetic phenomena. MNPs can thus exhibit many property changes with the reduction in size. Last (but no least), MNPs can show important macroscopic effects of interparticle magnetic interactions, which can involve additive forces (exchange), or attraction/repulsion (dipole).

Synthesis of MNPs

The most common method to synthesize MNPs are based on coprecipitation and microemulsion [11]. The coprecipitation method produces NPs by a pH change in a solution containing the desired metals in the form of nitrates or chlorides. Average size and size distribution, as well as shape depend on the pH and the ionic strength of the precipitating solution. In the microemulsion method, an aqueous metal solution phase is dispersed (entrapped) as microdroplets in a continuous oil phase within a micellar assembly of stabilizing surfactants. The advantage is that the microdroplets provide a confined space which limits the growth and agglomeration of NPs.

An emerging method for preparation of uniform NPs is the polyol technique, where metallic salts (acetates, oxalates), dissolved in an alcohol (such as diethylenglycol) are directly precipitated by high temperature decomposition [12]. This method can produce metals; by addition of a controlled amount of water, it can lead to oxide MNPs.

Spray and laser pyrolysis, with great commercial scale-up potential have been reported [13]. In spray pyrolysis, a solution of a ferric salt (and a reducing agent) is sprayed through a reactor to produce evaporation of the solvent within each droplet. In laser pyrolysis, the laser energy is used to heat a flowing mixture of gases leading to a chemical reaction. Under the appropriate conditions, homogeneous nucleation occurs and NPs are produced.

A different method utilizes high-energy ultrasound waves to create acoustic cavitations resulting in extremely hot spots. The sound waves produced by these cavities can lead to particle size reduction and hence the formation of NPs [14]. Other methods are based on electrochemical deposition of metal in a cathode [11], and also the use of magnetotactic bacteria [15].

Microwave absorption in MNPs

In this brief review, some of the recent developments in microwave absorption in MNPs. The response of ferromagnetic resonance (FMR) of MNPs (under different conditions) is first reviewed. FMR results on consolidated materials by spark plasma sintering (SPS) techniques are included, as this method allows the preparation of nanostructured ferrites (grains under

100 nm in size) with high densities. The behavior of electron paramagnetic resonance (EPR) of some relevant magnetically disordered MNPs systems is also presented. We devote a part of this review to the emerging low field microwave absorption technique (LFMA), which is a non-resonant method providing valuable information on magnetically-ordered materials. Results on MNPs in different aggregation states, as well as SPS-sintered materials are briefly reviewed.

As FMR [16] and EPR [17] techniques are well known, so no additional treatment of them is included here, except for some references.

2. Ferromagnetic resonance (FMR) in ferrite nanoparticles

Temperature and size dependences

One of the most studied phenomena on MNP's is the change of their FMR spectra as a function of temperature. The main parameters describing FMR signal are plotted versus temperature and eventually compared with the bulk counterpart, as an attempt to characterize the changes associated with the nanometric scale. Such parameters are generally peak-to-peak resonance linewidth, ΔH_{PP}, resonance field, H_{res}, and the intensity or height of the resonant absorption signal. Often, a simple linear dependence with T is observed [18-23]. Magnetic and structural phase transitions appear as a discontinuous event on this dependence [24].

The resonance field behavior with temperature for bulk and MNP's decreases with decreasing temperature, as a consequence of the enhancement of the contributions to the internal field associated with magnetic ordering (mainly exchange and anisotropy). This effect is stronger for small particles (figure 1) [25], revealing additional contributions to the internal field at low temperatures as a consequence of the size decreasing. These contributions can be assumed as an extra unidirectional internal field arising from surface disorder, where the magnetization processes are presumably to be isotropic and causes an extra shift of the resonance field. Therefore, as the surface area increases with decreasing the particle size, the isotropic effects on the magnetic resonance behavior are more pronounced and an additional distribution of energy barriers, promoted by surface isotropic disorder, must be assumed.

As observed in many works [23,25-27], the FMR spectra for MNPs at intermediate temperatures results to be a mixing of two lines: a broad component corresponding to typical anisotropic contributions and a narrow one, presumably corresponding to the surface isotropic contributions. This leads to a characteristic FMR shape for nanoparticulated systems. The general features include a broad component (becoming wider and shifting to lower fields upon cooling, see figure 2), a narrow component, and a large broadening and shifting as the particle size decreases. Figure 2 shows the FMR signal evolution with temperature variation from room temperature down to 100 K for a well diluted magnetite suspension [26]. A double component spectrum is well observed at high temperatures, while its two components seem to overlap in a single broad signal as temperature goes down. This can be attributed to an important decrease on isotropic contributions which causes the narrow component to disappear at low temperatures. In addition, decreasing temperature makes the broad component to widen, becoming more symmetric, and shifting to lower fields, thus revealing a random distribution of the anisotropy axis enhanced at low temperatures. Inter-particle interactions are negligible and do not contribute to this broadening since the high dilution of the studied suspension promotes isolation between particles.

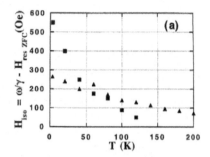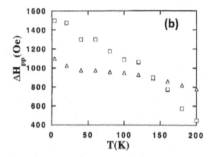

Figure 1. (a) Isotropic shift from the reference field (H_{res}=3.3 kOe) as a function of temperature for ZFC non-interacting maghemite (γ-Fe$_2$O$_3$) nanoparticle ferrofluid samples: (■) d = 4.8 nm (▲) d = 10 nm. (b) Peak to peak linewidth ΔH_{pp} as a function of temperature for ZFC non-interacting maghemite (γ-Fe$_2$O$_3$) nanoparticle ferrofluid samples: (□) d = 4.8 nm (Δ) d = 10 nm [25].

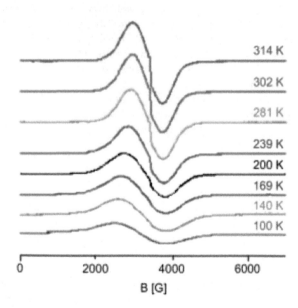

Figure 2. FMR signal evolution with temperature for a 0.1 wt% polymer Fe$_3$O$_4$ suspension [26].

Figure 3 shows the absorption signals for different particle distribution sizes of non-interacting maghemite (γ-Fe$_2$O$_3$) nanoparticle ferrofluid samples at room temperature [27]. The absorption signal changes drastically with changes on the particle size. Two limit cases are observed. The upper limit, for the largest size particles (sample 2, d = 10 nm), corresponds to a wide line shifted to a lower field in comparison with the reference (ω/γ = 3.3 kOe, where γ is the gyromagnetic ratio for free electrons). This line seems to be characteristic for anisotropic contributions. On the contrary, for the smaller size particles (sample 6, d = 10 nm) the spectrum shows a much narrower line, apparently characteristic for isotropic contributions and not shifted from the reference. The FMR signals for intermediate cases result in a more complex double-feature shape spectrum (samples 3, 4 and 5). Such behavior may be explained in terms of a mixing of two overlapped signals, the broad anisotropic shifted line and the narrower one, isotropic and not shifted. The coexistence of these two different contributions reveals the presence of a core-shell structure for the studied nanoparticles.

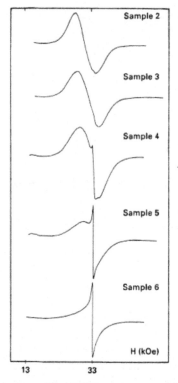

Figure 3. Influence of the particle size on X-band FMR spectra measured at room temperature for ZFC non-interacting maghemite (γ-Fe$_2$O3) nanoparticle ferrofluid samples. The particle size decreases from sample 2 to 6; sample 2 has the largest particle size distribution (d = 10 nm) and sample 6 the smaller particle size distribution (d = 4.8 nm) [27].

Particle concentration dependence

Double component FMR spectra have also been reported for solid and liquid Fe_3O_4 suspensions [26]. Important changes were observed on the absorption signal depending on the particle concentration. Figure 4 shows the decrease of the narrow component as the concentration increases. Both components exhibit also a broadening as the concentration increases. This important dependence of the signal linewidth with concentration is due to an increase of particle dipolar interactions at mean distances and a consequence of aggregation [28-30].

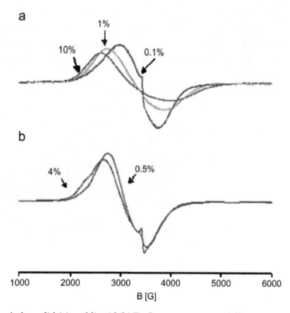

Figure 4. FMR signals for solid (a) and liquid (b) Fe_3O_4 suspensions at different concentrations [26].

Particle interactions always play an important role on the magnetic resonance absorption phenomena. FMR at different frequencies on dispersed KBr and then compressed $NiFe_2O_4$ commercial nanoparticles has recently been reported [31]. The resulting pellets showed strong shape anisotropy, as in-plane and out-of-plane analyzed measurements diverge with the increase of nanoparticle packing fraction (figure 5).

The resonance field decreases with increasing volume for in-plane measurements. In contrast, for out-of-plane measurements it increases with volume fraction. Both cases (in and out of plane), showed a resonance field shifted to a lower field in comparison with an ideal bulk as reported. As the packing fraction decreases, the in-plane and the out-of-plane curves converge to the same field value (≈ 0.04 T). This value can be identified as the average effective anisotropy field of the particles. A useful and powerful tool to estimate particle anisotropy can be based on these measurements.

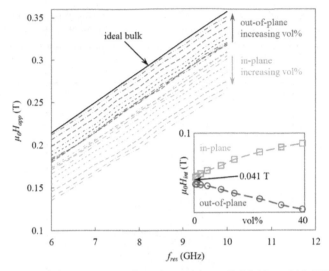

Figure 5. Frequency and shape orientation dependence of the applied field, at which FMR occurs, for commercial $NiFe_2O_4$ nanoparticles, dispersed in KBr and then compressed with different packing volume fractions. The inset shows the evolution of the resonance field as a function of the packing volume fraction [31].

Angular dependence

The anisotropy distributions and its FMR signal effects on MNP's can be carried out by measuring the angle variation between the cooling field of field-cooled (FC) samples. By comparing the obtained spectra for different orientations of FC samples, it is possible to determine the contribution of the anisotropy distribution within each particle to the FMR signal. The corresponding resonance lines often lead to a double component signal. It is then possible to separate the contributions which have a strong dependence with the angle from the weakly dependent counterpart. The narrow component can be attributed to surface anisotropy, while the the broad component should be associated with internal anisotropy of the NPs. Recent works have demonstrated the isotropic/anisotropic nature of most common MNP's, as described above. Figure 6 shows differences between FMR spectra for parallel and perpendicular configurations of FC samples. The samples consisted of maghemite nanoparticles with a particle diameter reported as 4.8 nm [27]. At the high temperature (top in figure 6), no variation as a function of θ is observed for the narrow component, while the broad component, on the contrary, shifts to higher resonance fields as the angle θ increases (bottom in figure 6). The narrow component is not affected by angular variations.

The resonance field of the anisotropic component generally is plotted against the angle θ, the dependences with θ use to be fitted, in good agreement, with $\sin^2 θ$ or $\cos^2 θ$ functions [24-32]. Its well established the axial symmetry nature for this kind of functions, as magnetization processes perform the higher energy absorption (energy improved by the resonance field) when the magnetizing field is perpendicular to the axial orientation, the resonance field also increases and finds a maximum at this position. The differences between the resonance field at θ = 0° and θ = 90°, are related to the anisotropy field, promoted by the respective axial anisotropy within the core of the nanoparticles revealing a mono-domain configuration on this region (figure 7).

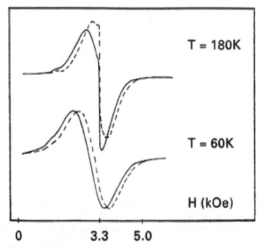

Figure 6. X-band FMR spectra of maghemite FC nanoparticles dispersed on glycerol, with a 10 kOe cooling field. Solid and dashed lines correspond respectively, to configurations with the scanning field parallel (θ = 0°) and perpendicular (θ = 90°) to the cooling field [27].

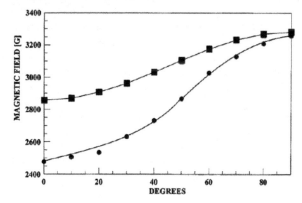

Figure 7. Dependence of the resonance field with θ for field-cooled (FC) hematite NPs at 300 and 155 K (lower curve) [24].

The fitting of the angular resonance field dependence is useful to determine anisotropy parameters. The dependence can be given by [32]:

$$H_r = H_o - H_a(1/2)(3cos^2\theta - 1) \tag{1}$$

where H_r is the resonance field, $H_a = 4|K|/M$ is the anisotropy field (with $|K|$ the absolute value of the anisotropy constant and M the magnetization), and H_o measures the g value. The g value is a constant describing the relation between the energy of the microwave radiation and the dc magnetic field. For the parallel configuration the resonance field is:

$$H_r = H_o - 4|K|/M \qquad (2)$$

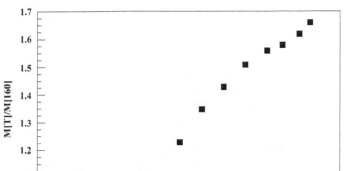

Figure 8. Reduced magnetization dependence with temperature for a FC hematite NPs [24].

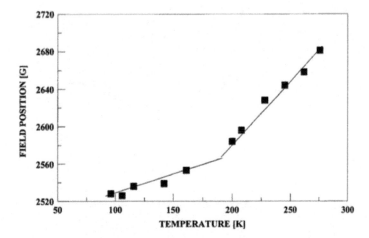

Figure 9. Resonance field dependence with temperature for hematite sample of Fig. 8 [24].

Magnetic phase transitions produce significant changes in $|K|/M$ and hence on the resonance field. The parallel configuration can therefore be used to investigate phase transitions.

Figure 8 shows the temperature dependence of magnetization of FC hematite NPs, parallel to the cooling field [24]. A change in the slope is observed in good agreement with observed transitions from a disordered phase to a magnetically ordered phase. In contrast with bulk hematite (antiferromagnetic above 260 K), these hematite NPs exhibited a superparamagnetic behavior, which can be explained in terms of a weakly ferromagnet below ~ 200 K, as shown in figure 8. The reported transition is also observed in plots of the resonance field against temperature (figure 9) and other FMR spectrum parameters, which suggest the FMR technique as a useful tool to investigate phase transitions on MNP's.

Effects of the aggregation state

Ferrite NPs can show the effects of two extreme aggregation states, namely monodisperse NPs, and clusters of a few hundreds of NPs. Samples of composition $Zn_{0.5}Ni_{0.5}Fe_2O_4$ were prepared by the polyol method, and designated as "A" for monodisperse state, and "B" for clusters. FMR spectra exhibited significant differences, as shown in Fig. 10. The decrease in the effects of surface for sample B appear in the form of a lower resonance field and an increase in the linewidth, as compared with sample A. The former can be understood in terms of a larger internal field in the cluster sample as a consequence of the aggregation of NPs; simply, the NPs inside the cluster tend to behave as grains in a bulk material. The effects of surface (crystal defects, unsatisfied bonds, etc.) are relieved by the presence of other NPs. Magnetic exchange interactions among neighboring NPs can also be assumed, which increases the internal field and thereby decreases the applied magnetic field needed to fulfill the Larmor equation conditions. On the other hand, these interactions increase the linewidth, especially the random distribution of anisotropy axes.

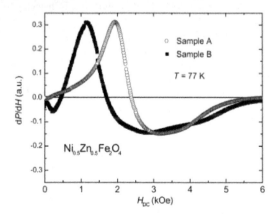

Figure 10. FMR spectra of Ni-Zn ferrites at 77 K, in monodisperse state (Sample A), and clusters (sample B) [18].

At room temperature, as shown in Fig. 11, isolated NPs show a superparamagnetic phase and again a larger resonance field with a significant reduction in linewidth as the factors just mentioned, associated with an ordered magnetic structure, are absent. The cluster sample exhibits a reduced linewidth, but always larger than the superparamagnetic state.

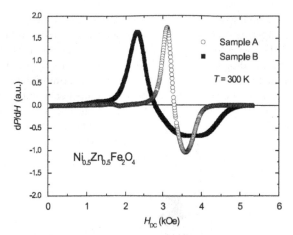

Figure 11. FMR spectra of Ni-Zn ferrites at 300 K, in monodisperse state (Sample A), and clusters (sample B) [18].

3. Ferromagnetic resonance in nanostructured ferrites

Ferrite NPs have to be consolidated as a high density solid for many applications (in electronic devices, for instance), where a powder is unstable. Typical sintering processes needing high temperatures are difficult to apply, as NPs tend to grow very rapidly at temperatures above 500°C, losing the nanometric size range and thus the different properties associated with this size range. A particularly well suited method to consolidate NPs into a high density nanostructured solid preserving grains within the nanometric range is Spark Plasma Sintering (SPS for short) [33]. Also known as pulsed electric current sintering (PECS), in this technique the sample (typically in the form of a powder) is placed in a graphite die and pressed by two punches at pressures in the 200 MPa range, while a strong electric current goes through the system; the die is shown in Fig. 12. SPS therefore consolidates powders under the simultaneous action of pressure and electric current pulses (typically of a few milliseconds in duration [33]).

The electric current pulses result in a very rapid heating of the sample, at rates as high as 1000°C/min. If the sample is a good conductor, current goes through it and the heating is even more efficient. A significant point is that the electric current has a significant impact on the atomic diffusion during the process [34]. The sintering process can then reach high densities at very low temperatures and extremely short times [35]. Obviously, SPS can also be utilized for reactive sintering involving a chemical reaction [36].

SPS has been used to consolidate spinel [37], garnet [38], and hexagonal ferrites [39]. In the case of spinel $Ni_{0.5}Zn_{0.5}Fe_2O_4$ ferrites, samples prepared in the form of 6-8 nm nanoparticles by the forced hydrolysis in a polyol method [12], were consolidated by SPS at temperatures in the 350-500°C range by times as short as 5 min. Just for comparison, the typical conditions for sintering in the classic solid state reaction are 1200°C for at least 4 hours. NPs growth was controlled, as the final grain size in the consolidated ferrite was about 60 nm, even for the highest (500°C) SPS temperature [40].The densities reached values as high as 94% of the theoretical value. FMR spectra obtained at 77 and 300 K are shown in Fig 13, together with the FMR signal corresponding to the original NPs. The resonance field exhibited a decrease as the SPS temperature increased, which can be explained in terms of the components of the total field in the Larmor expression, $\omega = \gamma H$. The main components of the total field are $H = H_{DC} + H_x + H_a + ...$, where H_{DC} is the applied field, H_x is the exchange field responsible for magnetic ordering, and H_a is the anisotropy field. As magnetic ordering sets in, H_x and H_a increase and therefore the applied field is decreased in order to fulfill the Larmor resonance conditions. As sintering progresses, the surface effects decrease and the material has the tendency to behave as a bulk ferrite.

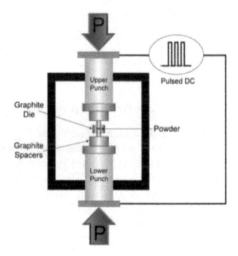

Figure 12. Schematic of spark plasma sintering apparatus [33].

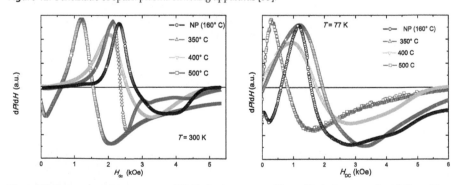

Figure 13. Ferromagnetic resonance of Ni-Zn ferrites prepared by a chimie douce method followed by SPS at temperatures in the 350-500°C range for 5 min. The FMR response of the original NPs is also shown for comparison. (Adapted from [40]).

All the samples exhibited a large broadening in the linewidth, generally interpreted by considering a random distribution of the anisotropy axis in single domain NPs [41]. The broadening decreases as consolidation increases, as the surface effects are diminished by formation of grain boundaries. At room temperature the linewidth decreases since all the samples approach the paramagnetic state where the internal field is eliminated and only the applied field is involved in the Larmor expression. The NPs signal exhibited the lowest linewidth as at room temperature these NPs are superparamagnetic; all the consolidated samples showed a ferrimagnetic behavior at 300 K as compared with the as-produced sample which has a blocking temperature about 90 K.

4. Paramagnetic resonance in ferrite nanoparticles

Zinc ferrite is a very good material to investigate the site occupancy by cations. In spite of a relatively large cation radius, Zn^{+2} has a tendency to occupy tetrahedral sites, while Fe^{3+} fill octahedral sites [1], in other words, a "normal" spinel. This arrangement leads to B-O-B superexchange interactions between iron cations and therefore to an antiferromagnetic structure with a Néel temperature about 9 K. In the case of nanosized Zn ferrite, however, the cation distribution can be significantly different; some degree of inversion occurs [1,42], with a fraction of Zn^{2+} on octahedral sites, and hence Fe^{3+} on both sites. The spinel becomes ferromagnetic, with a Curie temperature well above 9 K.

The variations in linewidth, ΔB_{PP}, and g-factor, g_{eff} in these ferrites depend mainly on the interparticle magnetic dipole–dipole interactions and intraparticle super-exchange interactions. On the other hand, the interparticle superexchange between magnetic ions (through oxygen) can reduce the value of linewidth. The magnitude of this interaction is determined by the relative position of metallic and oxygen ions. When the distance between the metallic and oxygen ions is short, the metal cations have an exactly half-filled orbital, and the angle between these two bonds is close to 180°, the superexchange interaction is the strongest [43].

Several factors can produce a distribution of local field, such as unresolved hyperfine structure, g-value anisotropy, strain distribution, crystal defects. The field strength on a particular spin is then modulated by local field distribution and leads to an additional linewidth broadening [44,45].

The strain distribution produced by a small average particle size can cause a small resonance signal [1]. As the particle size is increased, such small signal disappears and only the broad signal remains. This suggests that the particle size has a threshold value above which the strains are relieved.

In order to gain some insight into the relationship between internal structure and EPR spectra, it can be useful to compare the properties of several iron-based oxide NPs embedded in a polyethylene matrix, prepared by the same method: Fe_2O_3, $BaFe_2O_4$, and $BaFe_{12}O_{19}$ [46]. The experimental EPR spectra of the samples are presented in Figs. (14-20). At room temperature the EPR spectra of all the samples show a "two-line pattern" (Fig. 14) which is typical of superparamagnetic resonance (SPR) spectra. The relative intensity of these lines depends on the particle size and shape distribution function [47]. For the Fe_2O_3 and $BaFe_2O_4$ samples, the broad line predominates in the room temperature spectra; the opposite is observed for the $BaFe_{12}O_{19}$ sample, where the narrow line is more pronounced.

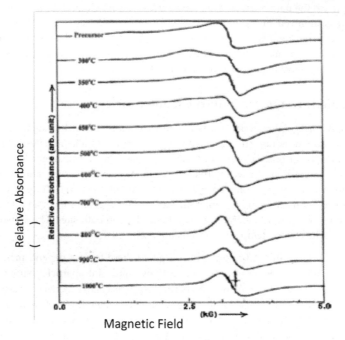

Figure 14. EPR spectra of zinc ferrite sintered at different temperatures.

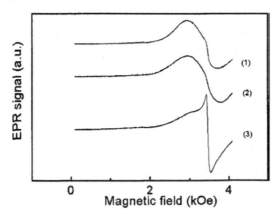

Figure 15. Room temperature EPR spectra of nanoparticles: Fe_2O_3 (curve1), $BaFe_2O_4$ (curve 2), and $BaFe_{12}O_{19}$ (curve3) [46].

At room temperature, the spectra of all samples show a "two-line pattern" (Figure 15) which is typical of superparamagnetic resonance. These spectra can be considered as a broader line superimposed on a narrow line. The relative intensity of these lines depends on the particle size and shape distribution function, as well as on the magnitude of the magnetic anisotropy. For Fe_2O_3 and $BaFe_2O_4$ samples, the broad line predominates in the RT spectra. This line is characterized by a peak-to-peak linewidth of $\Delta H \sim 850$ Oe and an effective g-value of 2.07. In the RT spectrum of $BaFe_{12}O_{19}$ sample, in contrast, the narrow line is more visible, with $\Delta H \sim 120$ Oe and $g \sim 2.0$.

At low temperatures, the EPR spectra of Fe_2O_3 change significantly (Fig.16). On cooling below 100 K, the broad line S1 shows a monotonous increase of the linewidth ΔH and a decrease of the amplitude A. However, below about 50 K new resonances, S2 and S3, appear in the spectra of Fe_2O_3 (Figs. 16 and 17). It is interesting that S2 behaves like a typical paramagnetic resonance signal, namely, the amplitude increases and the linewidth decreases as the temperature is diminished. The EPR spectra suggest that Fe_2O_3 nanoparticles contain both ferromagnetic and antiferromagnetic phases. The weak EPR signal of the rhombic symmetry ($g \approx 4.3$) that appears at low temperatures may be attributable to an α-Fe_2O_3 phase that undergoes an antiferromagnetic-like transition near 6 K. [46]

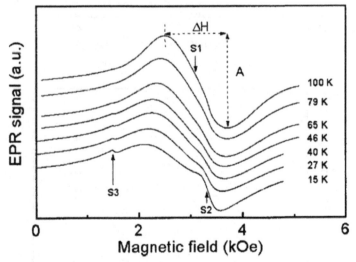

Figure 16. EPR spectra of Fe_2O_3 nanoparticles at different temperatures. [46]

EPR spectra of the $BaFe_2O_4$ nanoparticles at different temperatures are shown in Fig. 18. At all temperatures the spectra are broad and rather asymmetric. A significant shift of the line position to low magnetic fields and a marked spectrum broadening are observed at low temperatures.

On the other hand, the thermal variations of EPR spectra of the $BaFe_{12}O_{19}$ sample is typical of superparamagnetic resonance. The relatively narrow line that dominates at room temperature disappears as temperature decreases (Figs. 19 and 20). The BaFe$_{12}$O$_{19}$ nanoparticles reveal an EPR signal that is significantly narrowed at high temperatures by superparamagnetic fluctuations. This is evidence of the reduced magnetic anisotropy energy that may be due to the particle's nanosize (effective diameter <10 nm).

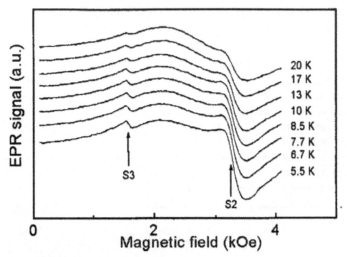

Figure 17. EPR spectra of Fe$_2$O$_3$ nanoparticles at low temperatures. [46]

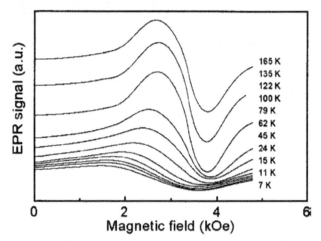

Figure 18. EPR spectra of the BaFe$_2$O$_4$ nanoparticles at different temperatures. [46]

These results show that EPR spectra at low temperatures are desirable for the correct identification of NPs and a comparison with high temperature experiments allows a better understanding of phenomena related with variations associated with nanosized materials.

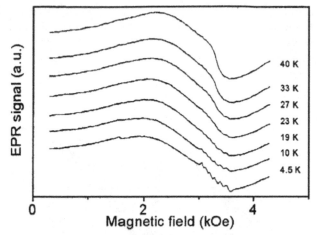

Figure 19. Low-temperature EPR spectra of BaFe$_{12}$O$_{19}$ nanoparticles. [46]

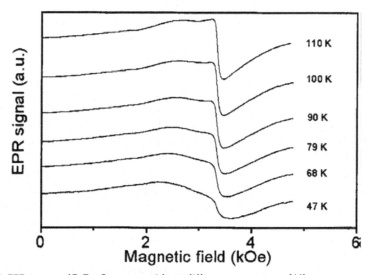

Figure 20. EPR spectra of BaFe$_{12}$O$_{19}$ nanoparticles at different temperatures. [46]

5. Low Field Microwave Absorption (LFMA)

Low field microwave absorption (LFMA for short) refers to the non-resonant, hysteretical losses of a material subjected to a high frequency electromagnetic field. Recently, it has become a useful method to investigate magnetization processes [48], magnetoelastic effects [49], phase transitions [50], non-aligned ferromagnetic resonance [51,52], spin arrangements [53,54]. LFMA is similar to giant magnetoimpedance (GMI) [55], but physically different to ferromagnetic resonance (FMR) [56]). GMI, generally defined as the variations of impedance of a magnetic conductor carrying an alternate electrical current when subjected to an external magnetic field [57], extends into a very wide frequency range. Clearly, GMI is a non-resonant phenomenon as confirmed by two facts: GMI does not fulfill the resonant Larmor conditions, and exhibits magnetic hysteresis.

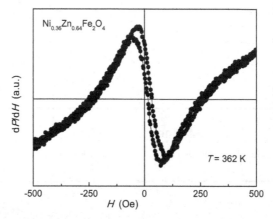

Figure 21. Typical "positive" LFMA signal from a Ni-Zn bulk ferrite (Adapted from [59]).

LFMA is associated with magnetization processes in magnetically ordered materials, in the process from the unmagnetized state to the magnetic saturation. In bulk ferro and ferrimagnetic materials, LFMA exhibits a flat response in the paramagnetic phase. To measure experimentally LFMA in a typical FMR/EPR facility, the applied field has to be cycled; usually between -1 kOe and +1kOe is enough. Also, a device to compensate for the remanent magnetization is needed in most electromagnets.

An important parameter is, of course, the total anisotropy field of the particular sample. In most cases, LFMA exhibits a critical behavior at the total anisotropy field in the form of a maximum and a minimum, leading to a characteristic signal as shown in Fig. 21 [59]. In bulk Ni-Zn ferrite, a correlation exists between the magnetocrystalline anisotropy and the half-peak-to-peak, measurement of LFMA. Figure 22 shows a comparison between the peak-to-peak LFMA field (divided by 2) [60] and a calculation from magnetocrystalline constant, K_1,published results [61]. The small differences can be attributed to shape anisotropy, as the calculation is based solely on K_1.

LFMA seems to be associated with spin structure. By convention, this signal can be assigned as "positive", simply because it has the same shape than the FMR/EPR signal, i.e., a maximum and a minimum when observed from left to right. A positive LFMA sign has been observed in most insulator and semiconductor materials, while a "negative" signal appears for most metallic conductors, as shown in Fig. 23, for a Co-rich CoFeBSi amorphous microwire [59]. An interesting result was found in bulk Ni-Zn ferrites showing the Yafet-Kittel triangular arrangement [62], by measuring the LFMA as a function of temperature in the 150 K-240K temperature range [63]. The LFMA sign changed from negative at ~ 154 K to positive at $T \geq 240$ K.

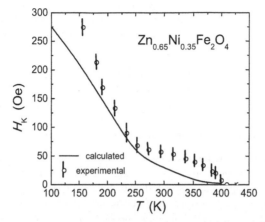

Figure 22. Fig. 21. Correlation between the field on LFMA critical points (divided by 2) and a direct calculation of magnetocrystalline anisotropy in $Zn_{0.65}Ni_{0.35}Fe_2O_4$ bulk ferrites [60]. K_1 anisotropy constant data for the calculation was obtained from ref [61].

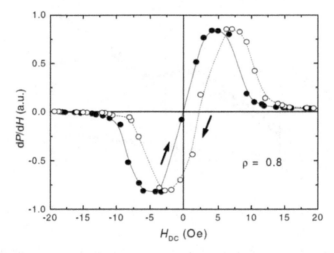

Figure 23. LFMA from a Co-rich CoFeBSi glass-covered amorphous microwire [59]. This is an example of a "negative" LFMA signal, mostly observed in conductor materials (compare with Fig. W above).

LFMA in NPs

LFMA can provide useful insights into the structure of NPs. When two different aggregation states are compared, i.e., monodisperse and clustered NPs with the same composition and NP diameter, clearly different spectra are obtained. By varying the synthesis conditions in the forced hydrolysis in a polyol method [12], Ni-Zn ferrites can be obtained as monodisperse, well crystallized ~ 6 nm NPs on one hand, and labeled as sample "A"; on the other, clusters about ~ 100 NPs constituted by NPs with the same composition and diameter [58], labeled as sample "B". It is interesting to mention that high resolution transmission electron microscopy (HRTEM) showed some epitaxial arrangements within the clusters of sample B.

Sample A showed a superparamagnetic behavior at room temperature, and a blocking temperature about 50 K [58]. Sample B, in contrast, exhibited a ferromagnetic behavior up to 300 K (blocking temperature > 300 K), in spite of being constituted by NPs of the same composition and NP diameter. Clusters effectively decrease the effects of surface producing samples with a general behavior between that of NPs and bulk materials.

Figure 24 shows the LFMA signal from sample A at 77 and 300 K. AT the low temperature, a positive LFMA behavior is observed, corresponding to a non-conductor material. At 300 K, however, a flat response appears, associated with the superparamagnetic phase of these monodisperse NPs. Clearly, the superparamagnetic phase behaves like a paramagnetic signal. This flat signal with a small slope is associated with the microwave absorption of non-interacting dipoles [64].

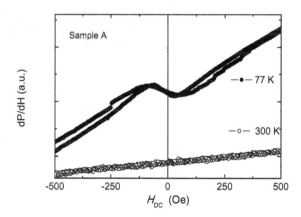

Figure 24. LFMA signal from monodisperse, ~ 6 nm diameter, $Zn_{0.5}Ni_{0.5}Fe_2O_4$ NPs, at 77 and 300 K (adapted from [58]).

Clusters of the same NPs, in contrast, showed a clear positive LFMA signal at room temperature, see Fig. 254. This is consistent, as this sample B behaved as a ferromagnetic phase close to the bulk properties.

For low temperatures, however, sample B exhibited a very different signal as shown in Fig. 26, with an important hysteresis and no similarity to either positive or negative character. As mentioned above, bulk Ni-Zn ferrites with Zn content $x > 0.5$ (in $Zn_xNi_{1-x}Fe_2O_4$) show a triangular Yafet-Kittel spin arrangement, which appears as an evolution from negative

LFMA signal at low temperatures towards a positive signal at temperatures close to room temperature [63]. Inspection of Figs. 24-26 reveals that in the case of sample A at 77 K (monodisperse NPs), no YK arrangement is manifested. Some evolution is found in sample B (clusters) at low T, but no full inversion of LFMA signal is observed. A preliminary explanation of these findings is that the surface effects inhibit the formation of the YK structure. Another possibility is based on the fact that cation distribution in NPs as synthesized by the polyol method is often different [42] as compared with bulk ferrites (prepared by the solid state reaction at high temperatures). A different cation distribution (in the present case a non-negligible concentration of Zn on B sites, for instance) could hence lead to different conditions for the formation of the YK structure.

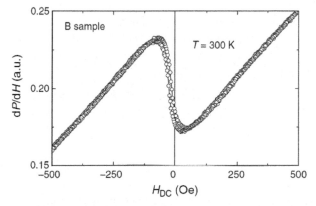

Figure 25. LFMA signal from epitaxial clusters of $Zn_{0.5}Ni_{0.5}Fe_2O_4$ ferrites, formed by a few hundreds of ~ 6 nm NPs, at 300 K.

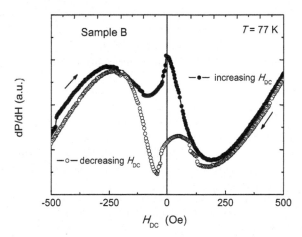

Figure 26. LFMA signal of epitaxial clusters of $Zn_{0.5}Ni_{0.5}Fe_2O_4$ ferrites, formed by a few hundreds of ~ 6 nm NPs, at 77 K.

6. Conclusions

Microwave absorption (MA) is a very sensitive phenomenon and has become an extremely powerful characterization tool. MA accurately depends on all the factors surrounding unpaired electrons; it can play a significant role in the characterization of the complex and fascinating development of magnetic nanoparticles. In this brief review, recent results on the characterization of magnetic nanoparticles and consolidated spinel ferrites by means of ferromagnetic resonance, paramagnetic resonance, and low field microwave absorption have been presented.

Author details

Gabriela Vázquez-Victorio, Ulises Acevedo-Salas and Raúl Valenzuela
Institute for Materials Research, National Autonomous University of México, México

Acknowledgement

Authors acknowledge partial support for this work from ANR-CONACyT grant 139292, as well as PAPIIT-UNAM grant IN141012.

7. References

[1] Valenzuela R. Magnetic Ceramics. Cambridge University Press (2005) (ISBN: 0-521-01843-9).

[2] Amiri S, Shokrollahi H. The role of cobalt ferrite magnetic nanoparticles in medical science. Materials Science and engineering C 2013: 33(1) 1-8.

[3] Mahmoudi M, Sant S, Wang B, Laurent S, Sen T. Superparamagnetic iron oxide nanoparticles (SPIONs): development, surface modification and applications. Advanced Drug Delivery Reviews 2011: 63(1) 24-46-

[4] Kashevsky BE, Agabekov BE, Kashevsky SB, Kekalo KA, Manina EY, Prokhorov IV, Ulashchik VS. Study of cobalt ferrite nanosuspensions for low frequency ferromagnetic hyperthermia. Particuology 2008; 6(5) 322-333.

[5] Kliza E, Strijkers GJ, Nicolay K. Multifunctional magnetic resonance imaging probes. Recent Results in Cancer Research 2013; 187 151-190.

[6] Xu Y, Wang E. Electrochemical biosensors based on magnetic micro/nanoparticles. Electrochimica Acta 2012; 84(1) 62-73.

[7] Dahawan SK, Singh K, Bakhshi AK, Ohlan A. Conducting polymer embedded with nanoferrite and titanium dioxide nanoparticles for microwave absorption. Synthetic Metals 2009, 159(21-22) 2259-2262.

[8] Liu R, Lal R. Nanoenhanced materials for reclamation of mine lands and other degrade soils: a review. Journal of Nanotechnology 2012; 461468-1-461468-18. doi: 10.1155/2012/461468.

[9] Hua M, Zhang S, Pan B, Zang W, Lv L, Zhang Q. Heavy metal removal from water/wastewater by nanosized metal oxides. Journal of Hazardous Materials 2012; 211-212 317-331.

[10] Hosokawa M, Nogi K, Naito M, Yokoyama T. Nanoparticle Technology Handbook (Elsevier, Amsterdam) 2007.

[11] Tartaj P, Morales MP, Gonzalez-Carreño V, Veintemillas-Verdaguer S, Serna CJ. Advances in magnetic nanoparticles for biotechnology applications. Journal of Magnetism and Magnetic Materials 2005; 290(1) 28-34.

[12] Beji Z., Ben Chaabane T., Smiri L.S., Ammar S., Fiévet F., Jouini N., and Grenèche J.M. Synthesis of nickel-zinc ferrite nanoparticles in polyol: morphological, structural and magnetic properties. physica status solidi (a) 2006; 203(3) 504-512.

[13] Tartaj P, Morales MP, Gonzalez-Carreño V, Veintemillas-Verdaguer S, Serna CJ. The preparation of magnetic nanoparticles for applications in biomedicine. Journal of Physics D: Applied Physics 2003; 36 R182.

[14] Dang F, Enomoto N, Hojo J, Enpuku K. A novel method to synthesize monodispersed magnetite nanoparticles. Chemical Letters 2008; 37(5) 530-531.

[15] Matsunaga T, Okamura Y, Tanaka T. Biotechnological application of nano-scale engineered bacterial magnetic particles. Journal of Materials Chemistry 2004; 14(14) 2099-2105.

[16] Guimaraes AP. Magnetism and magnetic resonance in solids. Wiley VCH (1998). (ISBN: 0471197742).

[17] Well JA, Bolton JR. Electron paramagnetic resonance: elementary theory and practical applications 2nd edition. John Wiley and Sons, 2007. (ISBN: 047175496X); Brustolon MR. Electron paramagnetic resonance: a practitioner toolkit, 1st edition. John Wiley and Sons (2009) (ISBN: 0470258829).

[18] Valenzuela R, Herbst F, Ammar S. Ferromagnetic resonance in Ni–Zn ferrite nanoparticles in different aggregation states. Journal of Magnetism and Magnetic Materials 2012; 324(21) 3398-3401.

[19] Guskos N, Zolnierkiewicz G ,Typek J, Guskos A, Czech Z. FMR study of γ-Fe2O3 agglomerated nanoparticles dispersed in glues. Reviews on Advanced Materials Science 2007;14 57-60.

[20] Thirupathi G, Singh R. Magnetic Properties of Zinc Ferrite Nanoparticles. Institute of Electrical and Electronics Engineers Transactions on Magnetics 2012; 48 3630-3633.

[21] De Biasi E, Lima E, Ramos C A, Butera A, Zysler R D. Effect of thermal fluctuations in FMR experiments in uniaxial magnetic nanoparticles: Blocked vs superparamagnetic regimes . Journal of Magnetism and Magnetic Materials 2013; 326(1) 138-146.

[22] Sobón M, Lipinski I E, Typek J, Guskos A. FMR Study of Carbon Coated Cobalt Nanoparticles Dispersed in a Paraffin Matrix. Solid State Phenomena 2007; 128 193-198.

[23] Shames A I, Rozenberg E, Sominski E, Gedanken A. Nanometer size effects on magnetic order in La12xCaxMnO3 (x = 0.5 and 0.6) manganites, probed by ferromagnetic resonance. Journal of Applied Physics 2012; 111 07D701-1-07D701-2.

[24] Owens F. Ferromagnetic resonance observation of a phase transition in magnetic field aligned Fe_2O_3 nanoparticles. Journal of Magnetism and Magnetic Materials 2009; 321(15) 2386-2391.

[25] Gazeau F, Bacri J C, Gendron F, Perzynski R, Raikher Yu L, Stepanov V I, Dubois E. Magnetic resonance of ferrite nanoparticles: Evidence of surface effects. Journal of Magnetism and Magnetic Materials 1998; 186(2) 175-187.

[26] Noginov M, Noginova N, Amponsah O, Bah R, Rakhimov R, Atsarkin V A. Magnetic resonance in iron oxide nanoparticles: Quantum features and effect of size. Journal of Magnetism and Magnetic Materials 2008; 320(18) 2228-2232.

[27] Gazeau F, Bacri J C, Gendron F, Perzynski R, Raikher Yu L, Stepanov V I, Dubois E. Magnetic resonance of nanoparticles in a ferrofluid: Evidence of thermofluctuational effects. Journal of Magnetism and Magnetic Materials 1999; 202(2-3) 535-546.

[28] Edelman I, Petrakovskaja E, Petrov D, Zharkov S, Khaibullin R, Nuzhdin V, Stepanov A. FMR and TEM studies of Co and Ni nanoparticles implanted in the SiO_2 matrix. Applied Magnetic Resonance 2011; 40 363-375.

[29] Martinez B, Obradors X, Balcells Ll, Rouanet A, Monty C. Low temperature surface spin-glass transition in gamma-Fe2O3 nanoparticles. Physical Review Letters 1998; 80 181-184.

[30] Winkler E, Zysler R D, Fiorani D. Surface and magnetic interaction effects in Mn3O4 nanoparticles. Physical Review B 2004; 70 174406-1-174406-5.

[31] Song H, Mulley S, Coussens N, Dhagat P, Jander V. Effect of packing fraction on ferromagnetic resonance in NiFe2O4 nanocomposites. Journal of Applied Physics 2012; 111 (07E348) 07E348_1 - 07E348_3.

[32] De Biasi R, Devezas T. Anisotropy field of small magnetic particles as measured by resonance. Journal of Applied Physics 1978; 49 2466-2470.

[33] Munir Z.A., Quach D.V. and Ohyanagi M. Electric current activation of sintering: a review of the pulsed electric current sintering process. Journal of the American Ceramic Society 2011; 94(1) 1-19.

[34] Anselmi-Tamburini U, Garay JE and Munir ZA. Fundamental investigation on the spark plasma sintering/synthesis process III. Current effect on reactivity. Materilas Science and Engineering A 2005; 407 24-30.

[35] Mizuguchi T, Guo S and Kagawa Y. Transmission electron microscopy characterization of spark plasma sintered ZrB2 ceramicc. Ceramics International 2010; 36(3) 943-946.

[36] Regaieg Y, Delaizir G, Herbst F, Sicard L, Monnier J, Montero D, Villeroy B, Ammar-Merah S, Cheikhrouhou A, Godart C, Koubaa M. Rapid solid state synthesis by spark plasma sintering and magnetic properties of LaMnO3 perovskite manganite. Materials Letters 2012; 80(1) 195-198.

[37] Valenzuela R, Ammar S, Nowak S, and Vázquez G. Low field microwave absorption in nanostructured ferrite ceramics consolidated by spark plasma sintering. Journal of Superconductivity and Novel Magnetism 2012; 25(7) 2389-2393.

[38] Gaudisson T, Acevedo U, Nowak S, Yaacoub N, Greneche JM, Ammar S, Valenzuela R. Combining soft chemistry and Spark Plasma Sintering to produce highly dense and finely grained soft ferrimagnetic $Y_3Fe_5O_{12}$ (YIG) ceramics. (to be published).

[39] Nakamura T, Okano Y, Tabuchi M, Takeuchi T. Synthesis of hexagonal ferrite via spark plasma sintering technique. Journal of the Japan Society of Powder and Powder Metallurgy 2001; 48(2) 166-169.

[40] Valenzuela R, Beji Z, Herbst F, and Ammar S. Ferromagnetic resonance behavior of spark plasma sintered Ni-Zn ferrite nanoparticles produced by a chemical route. Journal of Applied Physics 2011; 109(*) 07A329-1-07A329-3.

[41] Sukhov A, Usadel KD, and Nowak U. Ferromagnetic resonance in an ensemble of nanoparticles with randomly distributed anisotropy axes. Journal of Magnetism and Magnetic Materials 2008; 320(1) 31-33.

[42] Ammar S, Jouini N, Fièvet F, Beji Z, Smiri L, Moliné P, Danot M, Grenèche JM. Magnetic properties of zin-ferrite nanoparticles synthesized by hydrolysis in a polyol medium. Journal of Physics: Condensed Matter 2006; 18(39), 9055-9069.

[43] Sui Y, Xu DP, Zheng FL and Su WH. Electrons spin resonance study of $NiFe_2O_4$ nanosolids compacted under high pressure. Journal of Applied Physics 1996; 80 719-723.

[44] A. Abragam and B. Bleaney, *Electron Paramagnetic Resonance of Transition Ions* (Clarendon Press, Oxford,1970).

[45] C. P. Poole and H. A. Farach, *Relaxation in Magnetic Resonance* (Academic Press, London, 1971).

[46] Koksharov Yu. A, Pankratov D A., Gubin SP, Kosobudsky ID, Beltran M, Khodorkovsky Y and Tishin AM. Electron Paramagnetic Resonance of Ferrite Nanoparticles. Journal of Applied Physics 2001;89 2293-2298.

[47] Kliava J and Berger R. Size and Shape Distribution of Magnetic Nanoparticles in Disordered Systems: Computer Simulations of Superparamagnetic Resonance Spectra. Journal of Magnetism and Magnetic Materials 1999; 205 328-342.

[48] Valenzuela R, Alvarez G, Montiel H, Gutiérrez MP, Mata-Zamora ME, Barrón F., Sanchez AY, Betancourt I, Zamorano R. Characterization of magnetic materials by low-field microwave absorption techniques. Journal of Magnetism and Magnetic Materials 2008; 320(14) 1961-1965.

[49] Montiel H, Alvarez G, Gutiérrez MP, Zamorano R, and Valenzuela R. The effect of metal-to-glass ratio on the low field microwave absorption at 9.4 GHz of glass coated CoFeBSi microwires. IEEE Transactions on Magnetics 2006; 42(10) 3380-3382.

[50] Montiel H., Alvarez G., Gutiérrez M.P., Zamorano R., and Valenzuela R. Microwave absorption in Ni-Zn ferrites through the Curie transition. Journal of Alloys and Compounds 2004; 369(1) 141-143.

[51] Prinz G.A., Rado G.T. and Krebs J.J. Magnetic properties of single-crystal {110} iron films grown on GaAs by molecular beam epitaxy. Journal of Applied Physics 1982; 53 (3) 2087-2091.

[52] Gerhardter F, Li Yi, and Baberschke K. Temperature-dependent ferromagnetic-resonance study in ultrahigh vacuum: magnetic anisotropies of thin iron films. Physical Review B 1993; 47(17) 11204-11210.

[53] Valenzuela R. The temperature behavior of resonant and non-resonant microwave absorption in Ni-n ferrites. In *Electromagnetic Waves* / Book 1, InTech Open Access

Publisher, edited by Vitaliy Zhurbenko, pp. 387 - 402 (2011), ISBN: 978 – 953 – 307 – 304 – 0. Online June 24, 2011 at: http://www.intechopen.com/articles/show/title/the-temperature-behavior-of-resonant-and-non-resonant-microwave-absorption-in-ni-zn-ferrites.

[54] Alvarez G., Montiel H., Barron J.F., Gutiérrez M.P., Zamorano R. Yafet-Kittel-type magnetic ordering in $Ni_{0.35}Zn_{0.65}Fe_2O_4$ ferrite detected by magnetosensitive microwave absorption measurements. Journal of Magnetism and Magnetic Materials 2009; 322(3) 348-352.

[55] Montiel H, Alvarez G, Betancourt I, Zamorano R. and Valenzuela R. Correlation between low-field microwave absorption and magnetoimpedance in Co-based amorphous ribbons. Applied Physics Letters 2005 ; 86(7) 072503-1-072503-3.

[56] Barandiarán M., García-Arribas, A., and de Cos, D. Transition from quasistatic to ferromagnetic resonance regime in giant magnetoimpedance. Journal of Applied Physics 2006; 99 103904-1-103904-4.

[57] Knobel M, Kraus L, and Vázquez M. Magnetoimpedance. Handbook of Magnetic Materials, edited by K.H.J. Buschow, Elsevier Amsterdam 2003; 15 497-563.

[58] Valenzuela R., Ammar S., Herbst F., Ortega-Zempoalteca R. Low field microwave absorption in Ni-Zn ferrite nanoparticles in different aggregation states. Nanoscience and Nanotechnology Letters 2011; 3(4) 598-602.

[59] Valenzuela R., Montiel H., Alvarez G., and Zamorano R. Low-field non-resonant microwave absorption in glass-coated Co-rich microwires. Physica Status Solidi A 2009; 4, 652-655.

[60] Valenzuela R. (Unpublished results)

[61] Broese van Groenou A., Schulkes J.A., and Annis D.A. Magnetic anisotropy in some nickel zinc ferrites. Journal of Applied Physics 1967; 38(3), 1133 (1967).

[62] Yafet Y. and Kittel, C. Antiferromagnetic arrangements in ferrites. Physical Review 1952; 87(2) 290-294.

[63] Alvarez G., Montiel H., Barrón J.F., Gutiérrez M.P. Zamorano R. Yafet-Kittel-type magnetic ordering on NiZnFe2O4 ferrite detected by magnetosensitive microwave absorption measurements. Journal of Magnetism and Magnetic Materials 2010; 322(3) 348-352.

[64] Alvarez G., Font R., Portelles J., Zamorano R., and Valenzuela R. Microwave power absorption as a function of temperature and magnetic field in the ferroelectromagnet Pb(Fe1/2Nb)O₃. Journal of the Physics and Chemistry of Solids 2007; 68(7) 1436-1442.

Chapter 7

Dynamic and Rotatable Exchange Anisotropy in Fe/KNiF$_3$/FeF$_2$ Trilayers

Stefania Widuch, Robert L. Stamps, Danuta Skrzypek and Zbigniew Celinski

Additional information is available at the end of the chapter

1. Introduction

Exchange bias with natural antiferromagnets is typically investigated for systems with only a single magnetic interface [1-7]. Of these, one of the best understood in terms of atomic level spin configurations is the epitaxial Fe/FeF$_2$ film system. The present work explores exchange bias in a structure in which an epitaxially grown KNiF$_3$ film is used to separate the Fe and FeF$_2$. The uniaxial magnetic anisotropy of the KNiF$_3$ is much weaker than that of the FeF$_2$, and the corresponding exchange bias and exchange anisotropies are also different.

By changing the thickness of the KNiF$_3$, we show that it is possible to change the exchange anisotropy and bias from that of the FeF$_2$ to that of the KNiF$_3$. In the limit of very thick KNiF$_3$, we observe unidirectional, uniaxial and rotatable exchange anisotropies. Competing effects from the FeF$_2$ are observed as the KNiF$_3$ thickness is reduced, allowing us to probe magnetic lengthscales associated with spin ordering near the KNiF$_3$ interfaces.

We used ferromagnetic resonance (FMR) to obtain values for exchange anisotropies and bias. The FMR response is provided by the Fe in our Fe/KNiF$_3$/FeF$_2$ systems, and magnetic anisotropies and other parameters are obtained by fitting the raw data to well known resonance conditions for thin films. In this chapter we show that the fitted values reveal an exchange anisotropy that appears to be dependent on applied field orientation relative to the magnetic anisotropy symmetry axes. Throughout the remainder of the chapter we label this anisotropy H_{dyn}^{AFM}. The onset of H_{dyn}^{AFM} with temperature is similar to that of a rotatable anisotropy observed previously by us for KNiF$_3$ systems [8, 9], and we suggest that H_{dyn}^{AFM} in the present chapter consists of an isotropic 'rotatable' component H_{rot} in addition to an orientation dependent, unidirectional component H_E.

It is useful at this point to summarize the relevant magnetic properties of the antiferromagnets and their associated exchange bias phenomena. KNiF$_3$ is a Heisenberg antiferromagnet with cubic perovskite structure and Nèel temperature T_N=250K [10-13]. The FeF$_2$ has a rutile-type crystal structure and Nèel temperature T_N=79K [14-17]. KNiF$_3$ has a very small uniaxial anisotropy in comparison to FeF$_2$ [12-17].

Results of magnetometry and FMR studies of bilayered thin films of single crystal Fe(001) and either single crystal or polycrystalline KNiF$_3$ are given in references [8, 9]. The smallest lattice mismatch to the Fe is 1.2%, assuming that the Fe/KNiF$_3$ interface coincides at the (100) face of both materials. This good lattice match between the ferromagnetic and antiferromagnetic layers preserves the cubic structure of both.

The (100) plane of KNiF$_3$ should be compensated with both sublattices present in equal numbers. This should also be true for polycrystalline KNiF$_3$ on Fe since all possible growth planes are reasonably well lattice matched with Fe. The most striking feature observed was a blocking temperature in the range 50 K to 80 K. This temperature is much lower than the 250K Néel temperature expected for bulk KNiF$_3$. Particularly relevant for the present work was the observation (by FMR) of a rotatable anisotropy that was an order of magnitude larger than the exchange bias.

As noted earlier, the Fe/FeF$_2$ system has been particularly well studied. Nogués et al. [1] have shown that the exchange bias depends strongly on the spin structure at the interface and in particular, on the angle between the FM and AFM spins. Work by Fitzsimmons et al. [18-22] using polarized neutron reflectometry and magnetometry show that spin configurations at the FeF$_2$ interface differ from the bulk, and can be linked to exchange anisotropies and bias phenomena.

We have used MOKE microscopy to observe domains in a Fe/FeF$_2$ system to identify interface regions in which spin arrangements are distinct from either AFM or FM magnetic spin arrangements [23]. Results indicate that the crystallographic arrangement affects the value of the exchange bias but not the temperature dependence. Measurements of the temperature dependence of domain density in conjunction with coercivity field enhancement and exchange bias suggest that exchange bias and coercivity are different in origin in the sense that the unpinned magnetic moments at the AFM/FM interface are responsible for the enhancement of the coercivity field while pinned moments shifts the hysteresis loop.

In what follows we first discuss the sample structure and characterization in Section 2. This is followed in Section 3 by a discussion of reference FM/AFM1 and FM/AFM2 bilayers, and FM/AFM1/AFM2 trilayer results. In Section 4 we present our micromagnetic model and discuss the possibility of different thickness regimes. Results and discussion are summarized in Section 5.

2. Sample preparation and structural characterization

Samples were grown on GaAs(001) substrates using Molecular Beam Epitaxy. Wafers were annealed at elevated temperatures (~450^{0}C) and sputtered with Ar ions until 4x6 reconstruction on the surface of GaAs was clearly visible. The Fe was deposited using K-cells and the fluorides were grown using e-beam evaporation at room temperature. A 0.6 nm thick Fe seed layer was followed by a 75 nm thick Ag buffer layer. At this point the structure was annealed at 350°C for 12h. From RHEED we confirmed that annealing produced an average terrace size of ~13 nm. Next an Fe(001) layer was grown at room temperature followed by a KNiF₃ layer and topped with a 50 nm FeF₂ layer. Series with different KNiF₃ thicknesses between 0 and 90nm were prepared. The structures were capped with a 2.5 nm thick Au layer to protect samples during measurements in ambient conditions.

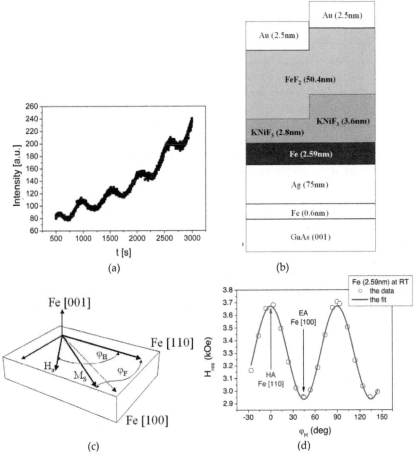

Figure 1. (a) Fragment of the RHEED intensity oscillations measured at the Fe specular spot during the growth; (b) a schematic diagram of the two grown Fe/KNiF₃/FeF₂ structures with 2.8nm- and 3.6nm-thick KNiF₃; (c) geometry for FMR measurement; (d) resonance field for different orientations of the static applied field relative to a magnetocrystalline anisotropy axis for Fe thin film (2.59nm). Unfilled circles denote the data taken at RT. Solid line exhibits the fit. Fe easy axis [100] (denoted as EA) for φH = 45° and Fe hard axis [110] (denoted as HA) are shown.

In figure 1(a), example RHEED intensity oscillations measured at the specular spot during the growth are shown, confirming layer by layer growth. Also, from RHEED we determined that the Fe layer is monocrystalline, and the fluorides layers are polycrystalline. Note that in order to reduce variations in growth conditions within the series, two thicknesses of KNiF₃ were grown on each Fe film, thereby forming two samples. The geometry is sketched in figure 1(b) for the 2.8 and 3.6 nm samples from the series. For each pair of samples the same roughness at the Fe/KNiF₃ interface is expected.

3. Experimental details and results

Ferromagnetic resonance was made at 24 GHz at temperatures between 24K and 300K, using the TE₀₁₁ mode of a cylindrical cavity. The temperature dependence measurements were carried out in a dewar equipped with a closed-cycle helium refrigerator. The temperature of the sample cavity was monitored with two E-type thermocouples. The dc signal was measured on a diode, and the first derivative of the power absorption signal with respect to the applied field was detected. A lock-in amplifier technique was used employing a weak, 0.5 Oe, 155 Hz ac modulation field superimposed on the applied dc magnetic field. The FMR absorption spectra were fit with standard Lorentzian function, providing directly the resonance field, H_{res}, and FMR linewidth, ΔH.

A sketch of the experimental geometry is shown in figure 1(c). In-plane angles for magnetization and applied field (φ_F and φ_H, respectively) are defined relative to the Fe[110] direction. Measurements were made by varying the angle φ_H. An applied magnetic field was swept between 1 – 6 kOe. These fields were large enough to saturate the samples to within 2^0 for all angles of applied field orientation. The samples were first measured at a room temperature, then were field cooled to 24K in the cooling field H_{cf} = 0.87kOe and H_{cf} = 2.07kOe for Fe easy axis (denoted as EA; $\varphi_H = 45^0$) and Fe hard axis (denoted as HA; $\varphi_H = 0^0$), respectively. These values of the cooling fields are significantly larger than field needed to saturate FM (~ 600 Oe along hard axis).

The effective magnetization $4\pi M_{eff}$, and a fourfold anisotropy field, $H_{||} = \dfrac{2K_{||}}{M_S}$ can be determined with FMR by measuring H_{res} for different orientations of the static applied field H_{app} relative to a magnetocrystalline anisotropy axis. An example for Fe thin film (2.59nm) is shown in figure 1(d). All FMR measurements were fit to the following resonance condition equation [8,9]:

$$H_{res} = \frac{1}{2}\left\{\left[\left(4\pi M_{eff} + \frac{3H_{||}}{4}\left(1 + \cos 4\phi_H\right)\right)^2 + \left(\frac{2\omega}{\gamma}\right)^2\right]^{1/2} - 4\pi M_{eff}\right\}$$
$$-H_{rot} + \frac{H_{||}}{8}\left(5\cos 4\phi_H - 3\right) - H_E \cos\left(\phi_H - \phi_E\right) \tag{1}$$

where the first term represents a constant shift in the resonance field baseline. This shift is due to the fact that $4\pi M_{eff}$ and $\dfrac{\omega}{\gamma}$ are typically an order of magnitude larger than Fe fourfold anisotropy. The second term consists of an induced shift in the position of the resonance line due to exchange coupling which has been attributed to rotatable anisotropy. The third and fourth terms contain the angular variation of H_{res} caused by the Fe fourfold and coupling induced unidirectional anisotropies, respectively. The easy direction of the unidirectional anisotropy is given by φ_E. Both terms H_{rot} and H_E are temperature dependent. As noted above, Eq. (1) is valid only for H_{res} large enough to saturate the magnetization.

A summary of results for effective magnetization and fourfold anisotropy fields determined from fits of H_{res} to Eq. (1) are summarized in table 1.

The fourfold anisotropy field of the 18 ML Fe films increased from ~440 Oe (average value for all measured samples) at room temperature to ~680 Oe at 24 K. For 36 ML thick Fe layers, a similar increase of the fourfold anisotropy field was observed (~540 Oe to ~710 Oe from room temperature to 24 K).

	Sample	RT		24 K	
		$4\pi M_{eff}$ (kOe)	$H_{\parallel}$ (kOe)	$4\pi M_{eff}$ (kOe)	$H_{\parallel}$ (kOe)
Fe	Fe(2.59nm)	15.85	0.41	16.17	0.61
	Fe(5.19nm)	18.91	0.53	18.65	0.68
Bilayers	Fe(2.60nm)/FeF$_2$(50nm)	14.15	0.44	12.88	0.68
	Fe(2.62nm)/KNiF$_3$(16nm)	14.98	0.45	14.04	0.67
	Fe(2.62nm)/KNiF$_3$(50nm)	15.17	0.43	14.50	0.69
Trilayers	Fe(2.60nm)/KNiF$_3$(0.8nm)/FeF$_2$(50nm)	14.18	0.43	12.94	0.65
	Fe(2.61nm)/KNiF$_3$(1.2nm)/FeF$_2$	15.92	0.48	14.79	0.71
	Fe(2.60nm)/KNiF$_3$(1.7nm)/FeF$_2$	14.44	0.47	13.15	0.76
	Fe(2.58nm)/KNiF$_3$(2.0nm)/FeF$_2$	15.62	0.43	15.15	0.645
	Fe(2.59nm)/KNiF$_3$(2.8nm)/FeF$_2$	14.76	0.45	13.84	0.66
	Fe(2.59nm)/KNiF$_3$(3.6nm)/FeF$_2$	14.90	0.46	13.76	0.68
	Fe(2.68nm)/KNiF$_3$(4.0nm)/FeF$_2$	15.62	0.42	15.52	0.63
	Fe(2.61nm)/KNiF$_3$(5.2nm)/FeF$_2$	15.51	0.45	14.08	0.77
	Fe(2.60nm)/KNiF$_3$(6.0nm)/FeF$_2$	15.00	0.49	13.37	0.77
	Fe(2.58nm)/KNiF$_3$(91.2nm)/FeF$_2$	17.27	0.43	14.89	0.64
	Fe(5.29nm)/KNiF$_3$(0.8nm)/FeF$_2$(50nm)	19.22	0.54	19.095	0.73
	Fe(5.29nm)/KNiF$_3$(2.8nm)/FeF$_2$	19.11	0.55	18.99	0.75
	Fe(5.15nm)/KNiF$_3$(4.0nm)/FeF$_2$	18.96	0.53	19.07	0.68

Table 1. Values of the effective demagnetizing field $4\pi M_{eff}$ (±0.06 kOe) and the fourfold iron anisotropy field $H_{\parallel}$ (±0.01 kOe) magnitudes at 24 K from linear fitting extrapolations for trilayer systems of Fe/KNiF$_3$/FeF$_2$(50nm) in comparison to single Fe layer and bilayers Fe/FeF$_2$, Fe/KNiF$_3$.

The effective demagnetizing field ($4\pi M_{eff}$) value is less than 21.4 kOe at room temperature. A possible explanation for this is the existence of surface anisotropy and uniaxial anisotropy fields, H_s and H_u respectively, such that $4\pi M_{eff} = 4\pi M_s - H_u - H_s$. It is interesting to note that the $4\pi M_{eff}$ for the thinner Fe layers decreases slightly with decreasing temperature from ~15.1 kG to ~14.1 kG between room temperature and 24 K. The thicker Fe layers show $4\pi M_{eff}$ nearly constant (~19.0 kG) between room temperature and 24 K. This is suggestive of an out of plane anisotropy induced by the KNiF₃ interface.

3.1. Fe/FeF₂ and Fe/KNiF₃ reference structures

The resonance field for the reference bilayers is less than that of the single Fe film data. This reduction is attributed to the AFM-induced dynamic anisotropy field H_{dyn}^{AFM}. The temperature dependence of the resonance field for easy and hard axis field orientation for bilayer Fe/FeF₂ is given in figure 2(a) and the inset shows the experimental results for the single Fe film. In figure 2(b) and (c) the temperature dependence of H_{dyn}^{AFM} for easy and hard axis field orientation for reference bilayers Fe/FeF₂ and Fe/KNiF₃ is presented. The values of H_{dyn}^{AFM} were obtained as the difference between the resonance field data of bilayers and the linear fit extrapolated to low temperature. For each sample the linear fit becomes from the high-temperature (i.e. greater than blocking temperature) resonance field dependence.

The values for the exchange bias H_E were obtained from scans with positive and negative applied magnetic fields. According to Eq. (1) the asymmetry between resonance fields at each 180^0 interval is $2H_E$. For Fe(2.60 nm)/FeF₂(50.1nm) structure value of H_E at 24K was found to be 185 Oe and is significally smaller (by a factor of four) than H_{dyn}^{AFM} determined for this sample (see figure2(b)). At the same time, the FMR results show a low value of H_{dyn}^{AFM} for Fe(2.62nm)/KNiF₃(16.0nm) bilayer (see figure2(c)). An exchange bias field of H_E = 19 Oe was found, in agreement with the value reported by Wee et al.[8, 9].

3.2. Fe/KNiF₃/FeF₂

An example of how H_{dyn}^{AFM} varies with temperature for trilayers is shown in figure 3. Values extracted for H_{dyn}^{AFM} display several interesting features. The first is that H_{dyn}^{AFM} is a non-monotonic function of field orientation. Figure 3 shows the temperature dependence of the H_{dyn}^{AFM} for Fe(2.6 nm)/KNiF₃(0.8 nm)/FeF₂(50 nm) structure for three different orientations of the applied field: along an easy axis (EA), a hard axis (HA) and midway between (IA; ϕ_H=22.5^0). Values of H_{dyn}^{AFM} are similar for hard and intermediate axis orientations and are significantly larger than that determined for easy axis orientation (at 24K: 1040 Oe, 1048 Oe and 757 Oe, respectively).

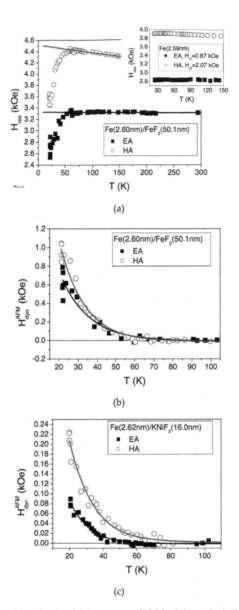

Figure 2. (a) Temperature dependence of the resonance field for bilayer Fe(2.60nm)/FeF$_2$(50.1nm). The inset in (a) shows the linear temperature dependence of the reference Fe layer, (b) Temperature dependence of H_{dyn}^{AFM} for bilayer Fe/FeF$_2$ and (c) Fe/KNiF$_3$. Filled squares denote measurements along Fe easy axis (EA; $\varphi_H = 45^0$) and unfilled circles along Fe hard axis (HA; $\varphi_H = 0^0$). The cooling field was H$_{cf}$ = 0.87kOe and H$_{cf}$ = 2.07kOe for Fe easy and hard axis, respectively. The lines are guides to the eye.

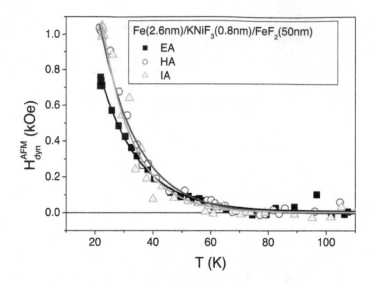

Figure 3. Temperature dependence of the H_{dyn}^{AFM} for trilayer structure Fe(2.6 nm)/KNiF₃(0.8 nm)/FeF₂(50 nm) for three different orientations of the applied field: along an easy axis (EA), hard axis (HA) and midway between (IA; ϕ_H=22.5⁰).

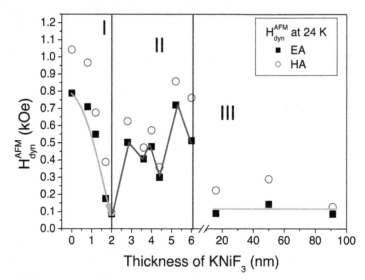

Figure 4. H_{dyn}^{AFM} as a function of the KNiF₃ film thickness at 24 K for structures with a 2.6 nm-thick Fe layer.

Another interesting feature is associated with the dependence of H_{dyn}^{AFM} on KNiF₃ thickness. In figure 4 we show H_{dyn}^{AFM} as a function of the KNiF₃ film thickness at 24 K for structures with a 2.6 nm-thick Fe layer. Note that the value of this anisotropy initially decreases very rapidly to almost zero for the 2nm-thick KNiF₃ layer. For samples with a thicker KNiF₃ layers (> 2 nm), the value of H_{dyn}^{AFM} increases and exhibits a non-monotonic dependence on thickness. For a fairly thick KNiF₃ layer (91.2nm) the value of the AFM-induced dynamic anisotropy reaches saturation at approximately 100 Oe. Note that even in this regime a difference of 40 Oe persists for values with the field measured along easy and hard axes.

We also mention results from field cooling. It has repeatedly been shown that the value of the unidirectional anisotropy H_E depends upon the cooling field strength. A stronger cooling field leads to more pinned spins and larger values for the exchange bias. We have tested the dependence of H_{dyn}^{AFM} on cooling field strength and determined that there is no such dependence in fields up to 1T. Instead, the magnitude of H_{dyn}^{AFM} for different orientations probably depends on magnitude of the AFM anisotropy field.

Finally, we demonstrate that H_{dyn}^{AFM} depends on the KNiF₃ thickness. Values of H_{dyn}^{AFM} are tabulated in table 2 for two different thicknesses of Fe in Fe/KNiF₃/FeF₂.

	H_{dyn}^{AFM} (t_Fe=2.6nm)		H_{dyn}^{AFM} (t_Fe=5.29nm)		$\dfrac{H_{dyn}^{AFM}(thin Fe)}{H_{dyn}^{AFM}(thick Fe)}$	
	EA (Oe)	HA (Oe)	EA (Oe)	HA (Oe)	along EA	along HA
Fe/KNiF₃(0.8nm)/FeF₂	706	966	301	429	2.34	2.25
Fe/KNiF₃(2.8nm)/FeF₂	502	625	244	333	2.06	1.88

Table 2. Comparison of the H_{dyn}^{AFM} values obtained for trilayer structures with two different thicknesses of FM layer.

Doubling the thickness of the Fe reduces H_{dyn}^{AFM} by approximately a factor of two for each KNiF₃ thicknesses, as one would expect for an interface effect. We therefore conclude that the results displayed in figure 4 may be related to spin structure *within* the KNiF₃.

3.3. FMR linewidth

Ferromagnetic resonance linewidth is the primary parameter, outside resonance field, considered in the experiment. In Figure 5(a,b,c) is shown FMR linewidth ΔH as function of temperature for selected reference Fe/FeF₂ and Fe/KNiF₃ samples and Fe/KNiF₃/FeF₂ trilayers.

The FMR linewidth for the single Fe layer displays a linear dependence on temperature. On the contrary, the observed temperature changes in linewidth of the exchange coupled systems are strong and non-linear. As an instance, ΔH of Fe(2.60nm)/KNiF₃(0.8nm)

/FeF$_2$(50.1nm) increases 12 times with decreasing temperature to 24K when the magnetic field was applied along an easy axis. When the FMR measurements were carried out with the field applied along hard axis, the observed increment is eightfold.

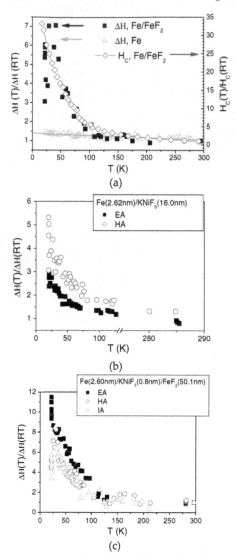

Figure 5. (a) Ferromagnetic resonance linewidths ΔH for Fe/FeF$_2$. For comparison, the temperature dependence of ΔH for single Fe is shown. Temperature dependence of coercive field Hc for Fe/FeF$_2$ is presented on the right-hand side scale. (b) Ferromagnetic resonance linewidths ΔH for Fe/KNiF$_3$ and (c) trilayer Fe/KNiF$_3$/FeF$_2$ as functions of temperature.

We proposed on explanation for FMR linewidth behavior that depends upon of inhomogeneity existing in the local fields acting on the ferromagnet. Some evidence for this may come from our MOKE magnetometer studies on Fe/FeF$_2$ structure [23] and the observations made by some of us for Fe/KNiF$_3$ system [8, 9]. For these reference bilayers a close correspondence between linewidth and coercive field Hc (see figure 5(a)) was noted. Similarly to FMR linewidth theory, the spatial inhomogeneities in the local internal fields can have a large effect on the coercive fields and magnetization loop widths.

4. Discussion

The temperature dependence of H_{dyn}^{AFM} described above behaves generally like that observed previously for Fe/KNiF$_3$ bilayers. A rotatable anisotropy was able to describe well FMR results in the previous work. Possible mechanisms underlying the formation of the rotatable anisotropy were discussed in references [8] and [9], and argued to be consistent with measured linewidths. The essential idea is that the Fe exchange couples to different regions of the KNiF$_3$ interface, and coupling between these regions can produce a rotatable anisotropy and coercivity as observed experimentally.

Similar arguments should apply in the present case. In the case of trilayers additional measurements were made in relation to reference bilayers, namely for Fe/KNiF$_3$(0.8nm)/FeF$_2$ sample temperature dependence of the H_{dyn}^{AFM} for three field orientations (EA, HA, IA) was studied. It turned out that the H_{dyn}^{AFM} is a non-monotonic function of the field orientation. We have at present no evidence to suggest a mechanism for this anisotropy, but can speculate that the presence of the second interface with high anisotropy FeF$_2$ introduces an additional competition that may affect alignment of regions in the KNiF$_3$.

It seems that a lot of information to verify the introduced model of H_{dyn}^{AFM} could be obtained from studies of Fe/AFM1/AFM2 with single crystalline AFMs. However, the studies of Fe/AFM bilayers, where AFM was deposited onto single crystal of Fe, either in the polycrystalline or single crystal form do not give a clear picture. The systematic study of the influence of in-plane crystalline quality of the antiferromagnet on anisotropies in Fe/FeF$_2$ structure were examined by Fitzsimmons et al [22]. In this study three types of samples were investigated with polycrystalline ferromagnetic Fe thin films and antiferromagnetic FeF$_2$ as: (i) untwinned single crystal; (ii) twinned single crystal, and (iii) textured polycrystal. The results obtained suggest that the value of exchange bias depends on many conditions, such as: the orientation between the spins in AFM and FM during field cooling; the choice of cooling field orientation relative to the AFM; engineering the AFM microstructure and so on. The bilayer Fe/AFM structures consisting of single-crystal Fe and KNiF$_3$, or KCoF$_3$, or KFeF$_3$ films were investigated by some of us [8,9,24,25]. All antiferromagnets in the samples had single crystalline structure or polycrystalline one with a high degree of texture. The crystalline quality of the antiferromagnets significantly affects the size of training effects, the magnitude of the parameters such as: FMR linewidth, the blocking temperature, the exchange bias. ΔHs of the samples with the single crystal fluorides were reduced compared to the polycrystalline structures. At the same time, the observed changes of the blocking temperature, the exchange bias, the coercivity vs the crystallity of the antiferromagnet film were different for different fluorides. It therefore seems indisputable that only the FMR linewidth is a parameter that uniquely altered depending on poly- or single crystalline nature of AFMs.

Perhaps the most curious feature observed was the complex dependence of H_{dyn}^{AFM} on KNiF₃

thickness as illustrated in figure 4. Three different behaviors were observed, corresponding to three thickness regions: 0 to 2 nm, 2 to 6nm and everything greater than 6 nm. The first two regions involve only a few monolayers of KNiF₃. Some insight into reasons underlying the existence of these three regions can be obtained by considering possible spin configurations in a model trilayer. The reason is that KNiF₃ is a weak antiferromagnet, and so we might expect that strong interlayer coupling to the Fe and FeF₂ might result in large modifications of the spin ordering near the interfaces. Characteristic length scales would correspond to a few nanometers. We explore this idea in some detail with a model described in the appendix. The model makes use of an iterative energy minimization scheme and was used to examine static equilibrium configurations of spins in a thin film trilayer using material parameters appropriate to the Fe/KNiF₃/FeF₂ trilayers studied in this work. Details of the model are left for the appendix, and we discuss below only the essential results and implications for understanding the dependence of H_{dyn}^{AFM} on KNiF₃

thickness.

According to our calculations, for example for trilayer structure Fe/KNiF₃(3ML)/FeF₂, we find that for angles θ_H (the direction of the applied field is characterized by the angle θ_H) from 0^0 to -95^0 the net moments in the first layers of both AFM's oppose the direction of the applied field. Between -95^0 and -96^0 a sudden change in spin orientation appears in the first layer of the FeF₂. In order to understand how these results may illuminate our measured results for H_{dyn}^{AFM}, we note first that the canted spin configuration calculated for all KNiF₃

thicknesses is suggestive of Koon's model [26] for exchange bias. In this model, a net moment at the interface is created by spin canting at a compensated interface of the antiferromagnet in contact with the ferromagnet. It was later pointed out by Schulthess and Butler [27] and discussed by Stiles and McMichael [28] that this canting was not itself sufficient to produce bias due to instabilities in the canted configuration. The instabilities allow the canted moment to reverse into a configuration whose energy is the same as the original field cooled orientation energy.

The same principle applies in our case for the largest thickness KNiF₃ films. However for small KNiF₃ thicknesses, the energy of the reversed state is not the same as for the field cooled orientation. The difference in energies for the $\theta_H=0$ and $\theta_H=180°$ configurations can be sizable, and is shown in figure 6 as a function of KNiF₃ thickness where 1ML=a(KNiF₃)=0.4013nm (Muller et al. [10]).

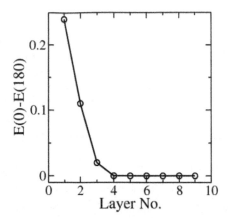

Figure 6. The difference in energies for the $\theta_H=0^\circ$ and $\theta_H=180^\circ$ configurations as a function of KNiF₃ thickness, where 1ML=a(KNiF₃)=0.4013nm (Muller et al.[10]).

This asymmetry in energies disappears when the KNiF₃ is thicker than 4 ML. This thickness is characteristic of the depth over which a twist can develop in the KNiF₃. Above 4 ML, the twist can extend to 180°, and behave like the 'partial' wall in the Stiles and McMichael picture of exchange bias[28, 29].

On the basis of this, we suggest that H_{dyn}^{AFM} is associated with the gradient of the energy as a function of θ_H, and that this energy is different for small and large θ_H for KNiF₃ thicknesses less than a domain wall width. KNiF₃ thicknesses sufficiently larger than this support formation of a partial wall that effectively isolates the Fe from the KNiF₃ interface. This results in an energy whose gradient is the same for 0 and 180° orientation of the field. As such, the H_{dyn}^{AFM} arises in this model as a kind of susceptibility of the spin configuration to the orientation of the Fe magnetization (which is controlled by the external applied field).

In this model for H_{dyn}^{AFM}, the thinnest KNiF₃ thickness region creates asymmetry in energies at 0 and 180° because the KNiF₃ is not wide enough to support a complete partial wall. In this region the effect of FeF₂ is apparent. The thickest KNiF₃ films support a complete partial wall, and the resulting H_{dyn}^{AFM} are associated with energy gradients that are symmetric with respect to reversal of the applied field. The intermediate thickness region allows spin configurations that are strongly thickness dependent with strongly distorted partial walls. The H_{dyn}^{AFM} in this region vary strongly with KNiF₃ thickness, in a manner determined by details of configurational energies of the distorted partial wall. Lastly, we note that this picture can also explain the dependence of H_{dyn}^{AFM} on field orientation with respect to easy and hard directions. This follows because the magnitude of H_{dyn}^{AFM} will depend on orientation of the Fe with respect to anisotropy symmetry axes.

5. Conclusions

In summary, we show that a ferromagnet exchange coupled to an antiferromagnetic bilayer can allow the character and strength of exchange anisotropy to be modified. We have studied using FMR exchange bias and exchange anisotropy for Fe exchange coupled to KNiF$_3$/FeF$_2$. The trilayer was grown by MBE. The temperature dependence of the ferromagnetic resonance peak position shows a characteristic negative shift of the resonance from that of a single Fe layer. This negative shift is a direct result of the exchange coupling between ferromagnetic and antiferromagnetic layers and results in a dynamically induced field H_{dyn}^{AFM} .

The value of H_{dyn}^{AFM} is different for measurements performed along ferromagnet easy and hard axes for all the bilayer and trilayer samples. The largest values were found for the field along Fe hard axis. The dependence of the H_{dyn}^{AFM} on the thickness of the KNiF$_3$ is non-monotonic, reaching a minimum for 2.0nm thick KNiF$_3$. For structures with thinner KNiF$_3$, the magnitude of the H_{dyn}^{AFM} increases with decreasing thickness, reaching the maximum expected for the bilayer Fe/FeF$_2$.

Results from a calculational model for the equilibrium configuration of spins in a ferromagnet/antiferromagnet/antiferromagn trilayer suggests that a form of spin canting may occur at the antiferromagnetic interfaces. For sufficiently thin KNiF$_3$, significant spin canting at the Fe and FeF$_2$ interfaces occurs due to exchange coupling. Effects of the FeF$_2$ are effectively isolated by KNiF$_3$ thick enough to support a partial magnetic domain wall. As a general result, we suggest that H_{dyn}^{AFM} is a measure of the susceptibility of the interface spin ordering to interface coupling in the KNiF$_3$.

Appendix

A numerical mean field model for equilibrium spin orientations in a ferromagnet/ antiferromagnet/antiferromagnet is described in this appendix. The spins are treated as classical vectors on a lattice.

The energy of a spin configuration is determined in the following way. An atomistic approach is employed in a mean field approximation. A cubic lattice of vector spins is considered consisting of N layers representing the first antiferromagnet (called AFM1 and representing the KNiF$_3$) and M layers representing a second antiferromagnet (called AFM2 and representing the FeF$_2$). The outermost layer of AFM1 is in contact with a block spin representing the FM. Each layer is described by a unit cell of spins representing the two antiferromagnetic sublattices. Periodic boundary conditions are assumed in the plane of each layer.

The ground state spin configuration is found as follows. The applied magnetic field is set, and a spin site in the FM/AFM1/AFM2 trilayer is chosen. The algorithm is begun by picking a set of values for the initial configuration of spins in each layer corresponding to a field cooled orientation. Exchange coupling between spins is taken into account in addition to anisotropy energies, so that the energy of the spins in any given layer depends on the orientation and magnitude of the spins in the nearby layers. The orientation of a spin in layer "i" is characterized by the angle θ_n made with respect to the applied field. A low energy magnetic configuration is obtained when the energy is minimized [30] i.e. the angle θ_n is determined by rotating the spin into the direction of a local effective field calculated from the gradient of the energy at the spin site. The procedure determines the configuration in the following manner: (i) a given layer is randomly chosen and the spins in that layer are rotated to be parallel to the local effective field; (ii) the process is continued until one has a self-consistent, stable state where all the spins are aligned with the local effective fields produced by the neighboring spins.

In this model, the multilayer is treated as an effective one-dimensional system with one spin representing an entire sublattice in a given layer. Only nearest-neighbor exchange coupling is considered. The energy is defined by:

$$H = H_{AFM} + H_{int} \tag{A.1}$$

$$H_{AFM} = H_a^{1,2} + H_b^{1,2} \tag{A.2}$$

where H_{AFM} is the Hamiltonian representing a given antiferromagnet with its both sublattices, and H_{int} is the Hamiltonian describing the interfaces contribution.

The geometry is defined in figure 7. The direction normal to the layer planes is y. Each xz plane, for a given value of y, depicts one layer of an AFM.

Both AFMs, KNiF₃ and FeF₂, exhibit uniaxial anisotropy with easy axes in the +z and −z directions. The first AFM is KNiF₃ and is known G-type meaning that the nearest neighbor coupling is antiferromagnetic. The nearest neighbor exchange constants for each AFM1 layer (J_{NNP}) and between AFM1 layers (J_{NN}) are defined as negative values. Each layer of the KNiF₃ is assumed to be compensated such that both AFM1 sublattices are present within a given layer. The second AFM is FeF₂ and is regarded as completely uncompensated within a given layer. In this case, the J_{NNP2} exchange coupling constant along x is defined as positive.

The complete Hamiltonian can be written as follows:

$$H_a^1 = -\mathbf{H}_{app} \cdot \mathbf{S}_{n(AFM1)}^a$$
$$+ \sum_{m=\pm 1} \left[J_{NN} \mathbf{S}_{n(AFM1)}^a \cdot \mathbf{S}_{(n+m)(AFM1)}^b + 4 J_{NNP} \mathbf{S}_{n(AFM1)}^a \cdot \mathbf{S}_{n(AFM1)}^b \right] \tag{A.3}$$
$$- K_u \left(\mathbf{S}_{n(AFM1)}^a \right)^2$$

$$H_b^1 = -\mathbf{H}_{app} \cdot \mathbf{S}_{n(AFM1)}^b$$
$$+ \sum_{m=\pm 1} \left[J_{NN} \mathbf{S}_{n(AFM1)}^b \cdot \mathbf{S}_{(n+m)(AFM1)}^b + 4 J_{NNP} \mathbf{S}_{n(AFM1)}^b \cdot \mathbf{S}_{n(AFM1)}^a \right] \tag{A.4}$$
$$- K_u \left(\mathbf{S}_{n(AFM1)}^b \right)^2$$

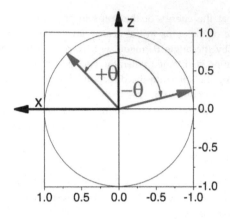

(a)

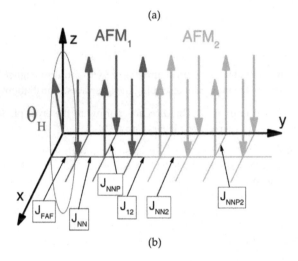

(b)

Figure 7. Geometry of the system (a) defining an angle of the applied field within the film plane θ_H, (b) the direction normal to the layer planes is y. Each xz plane, for a given value of y, depicts one layer of an AFM.

Similar terms can be written for the second antiferromagnet AFM2. Here, however, a given AFM2 plane is regarded, as completely compensated and only one sublattice is present in the interface:

$$H_a^2 = -\mathbf{H}_{app} \cdot \mathbf{S}_{n(AFM2)}^a$$
$$+ \sum_{m=\pm 1} J_{NN2} \mathbf{S}_{n(AFM2)}^a \cdot \mathbf{S}_{(n+m)(AFM2)}^b + \sum_{n'} J_{NN2P} \mathbf{S}_{n(AFM2)}^a \cdot \mathbf{S}_{n'(AFM2)}^a \qquad (A.5)$$
$$-K_{u2} \left(\mathbf{S}_{n(AFM2)}^a \right)^2$$

$$H_b^2 = -\mathbf{H}_{app} \cdot \mathbf{S}_{n(AFM2)}^b$$
$$+ \sum_{m=\pm 1} J_{NN2} \mathbf{S}_{n(AFM2)}^b \cdot \mathbf{S}_{(n+m)(AFM2)}^a + \sum_{n'} J_{NN2P} \mathbf{S}_{n(AFM2)}^b \cdot \mathbf{S}_{n'(AFM2)}^b \qquad (A.6)$$
$$-K_{u2} \left(\mathbf{S}_{n(AFM2)}^b \right)^2$$

Finally, the interface term with the FM is given by:

$$H_{\text{int}} = J_{FAF} \mathbf{S}_{FM} \cdot \mathbf{S}_{n=1(AFM1)}^a - J_{FAF} \mathbf{S}_{FM} \cdot \mathbf{S}_{n=1(AFM1)}^b$$
$$+ J_{12} \mathbf{S}_{n=\max(AFM1)}^a \cdot \mathbf{S}_{n=1(AFM2)}^a - J_{12} \mathbf{S}_{n=\max(AFM1)}^b \cdot \mathbf{S}_{n=1(AFM2)}^a \qquad (A.7)$$

The notation is defined by:

- H_{app} is the applied magnetic field,
- $\mathbf{S}_n^a, \mathbf{S}_n^b$ are the spin vectors at layer "n" for sublattice "a" and "b" of the AFM1 and AFM2, respectively. No out-of-plane spin component is considered within this model. The only spin arrangement is to lie within xz plane i.e. $(x, 0, z)$,
- $\mathbf{S}_{n'}^a, \mathbf{S}_{n'}^b$ are the spin vectors at layer "n" for sublattice "a" and "b" , respectively of the AFM2. Here, the different spin orientation is allowed within one n-th layer but again, no out-of-plane spin component is allowed to appear within this model,
- K_u, K_{u2} are site anisotropies for spins at layer "n" of AFM1 and AFM2, respectively.

The exchange coupling constants are defined as:

- J_{NN}, J_{NN2} are the exchange interaction constants between spins at "n" and "$n+m$" layers of AFM1 and AFM2, respectively,
- J_{NNP}, J_{NNP2} are the exchange interaction constants within n-th layer of AFM1 and AFM2, respectively,
- J_{FAF} is the interface exchange coupling constant between the FM layer and the first antiferromagnet,
- J_{12} is the interface exchange coupling constant between the first and the second AFM1/AFM2 layer.

Note that the sign of these exchange interactions has been introduced through the numerical values of the parameters. All the values of anisotropy and exchange coupling are given in field units and are summarized in table 3. All calculations were made for 50 atomic layers in AFM2.

An example arrangement of spins in the AFM1 and AFM2 is depicted in figure A1. Bulk values of the anisotropy and the exchange coupling constants for KNiF$_3$ and FeF$_2$ have been used in all calculations. The interface exchange coupling constant between AFMs (J_{12}) have been assumed to be a geometrical mean of the exchange coupling constants of the adjacent layers. The interface exchange coupling between FM and AFM1 is defined as ZJ_{NN}, i.e. the exchange coupling within the first AFM1 multiplied by the number of nearest neighbors of each Fe spin site.

AFM	Exchange field (kOe)	Anisotropy field (kOe)	*anisotropy field* / *exchange field*
AFM1: KNiF$_3$	$H_{ex} = 3500$[a]	$H_{Ku} = 0.080$[b]	$2.4 \cdot 10^{-5}$
AFM2: FeF$_2$	$H_{ex2} = 434$[c]	$H_{Ku2} = 149$[c]	0.33[d]

[a] M.E. Lines, Phys. Rev. **164**, 736 (1967).
[b] H. Yamaguchi, K. Katsumata, M, Hagiwara, M. Tokunaga, H. L. Liu, A Zibold, D. B. Tanner, and Y.J.Wang, Phys. Rev. B **59**, 6021 (1999).
[c] M.L. Silva, A.L. Dantas, and A.S. Carrico, Solid State Commun. **135**, 769 (2005).
[d] D.P. Belanger, P. Nordblad, A.R. King, V. Jaccarino, L. Lundgren, and O.Beckman, J. Magn. Magn. Mater. **31-34**, 1095 (1983).

Table 3. Used bulk values of the exchange and anisotropy fields for the KNiF$_3$ and FeF$_2$.

After converging to a stable configuration, the total energy per spin site was calculated as a function of the angle of the applied magnetic field, θ_H. Example results are shown in figure 8(a) for different thicknesses of the first AFM1.

Note that the horizontal axis (θ_H (deg)) is the same for all calculated structures, but the total calculated energy is not to scale in order to visibly depict the calculated results for each thickness of AFM1. A discontinuity in the total energy per spin site appears for thicknesses of AFM1 ranging from 1ML to 9ML. The values of angle θ_H for which the discontinuity appears (θ_{Hdis}) is presented in Fig 8(b). Three regions can be distinguished in these results: 1ML to 4ML, 5ML to 9ML, and > 9ML.

The different regions correspond to different spin arrangements within each AFM layer. An example for the 3ML thick AFM1 is shown in figure 9. The spin arrangements are presented in terms of the angle that each spin vector makes with the z-direction. Here, blue arrow represents the applied field, and red arrows are used for the spins in the three layers of AFM1 and green arrows represent the first few spin sites of AFM2.

The comparison between a spin arrangement for $\theta_H=-95^0$ (top diagrams) and for $\theta_H=-96^0$ (bottom diagrams) is shown. The central diagram in figure 9 shows the spin arrangement when the field is applied along +z direction. The first layer of AFM1 experiences spin-canting. An energy minimum for the fully compensated interface has been obtained for perpendicular interfacial coupling between the FM and AFM spins. The spin-canting and the resulting net moment induced within the first layer of the AFM1 always opposes the direction of the applied field. Moreover, spin-canting appears also in the second antiferromagnet AFM2 for the first two layers. It is important to note that the direction of the net moment induced in the first layer of AFM2 does not correlate directly to the direction of the applied magnetic field.

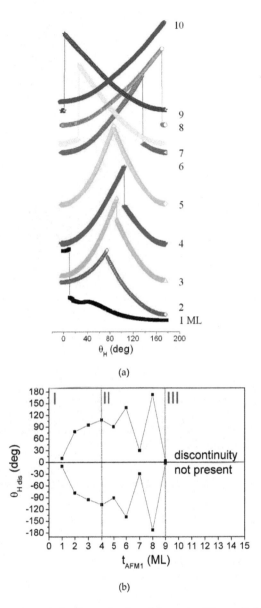

(a)

(b)

Figure 8. (a) The total energy per spin site as a function of the angle of the applied magnetic field θ_H for different thicknesses of the first AFM1. The horizontal axis (θ_H (deg)) is the same for all calculated structures, but the total calculated energy is not to scale in order to visibly depict the calculated results for each thickness of AFM1, (b) The values of angle θ_H for which the discontinuity appears (θ_{Hdis}) for thicknesses of AFM1 ranging from 1ML to 9ML. The lines are guides to the eye.

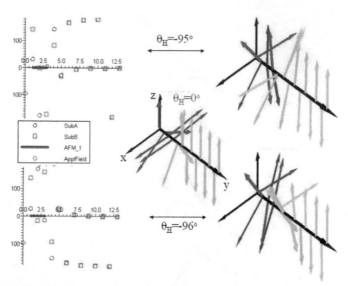

Figure 9. The spin arrangements are presented in terms of the angle that each spin vector makes with the z-direction for the 3ML-thick AFM1. Blue represents the applied field H$_a$, red arrows are used for the spins in the three layers of AFM1 and green represent the first few spin sites of AFM2.

Author details

Stefania Widuch
Center for Magnetism and Magnetic Nanostructures, University of Colorado, Colorado Springs, Colorado, USA
School of Physics (M013), University of Western Australia (UWA), Crawley, WA, Australia
Institute of Physics, University of Silesia, Katowice, Poland

Robert L. Stamps
School of Physics and Astronomy, Kelvin Building, University of Glasgow, Scotland, UK

Danuta Skrzypek
Institute of Physics, University of Silesia, Katowice, Poland

Zbigniew Celinski
Center for Magnetism and Magnetic Nanostructures, University of Colorado, Colorado Springs, Colorado, USA

Acknowledgement

RLS acknowledges the Australian Research Council. The work at UCCS was supported by the National Science Foundation Grants (DMR 0605629 and DMR 0907053). SW acknowledges funding through the IEEE for the visit at the UWA Perth.

6. References

[1]   J. Nogués, and I. K. Schuller, J. Magn. Magn. Mater. 192, 203 (1999).

[2]   A.E. Berkowitz, and K. Takano, J. Magn. Magn. Mater. 200, 552 (1999).

[3]   R.L. Stamps, J. Phys. D: Appl. Phys. 33, R247 (2000).

[4]   M. Kiwi, J. Magn. Magn. Mater. 234, 584 (2001).

[5]   J. Nogués, J. Sort, V. Langlais, V. Skumryev, S. Suriñach, J.S. Muñoz, and M.D. Baró, Phys. Rep. 422, 65 (2005).

[6]   A.E. Berkowitz, and R.H. Kodama, in *Contemporary Concepts of Condensed Matter Science*, editted by D.L. Mills and J.A.C. Bland (Elsevier B.V. 2006), Chapter 5, p. 115.

[7]   J. Stöhr, and H.C. Siegmann, *Magnetism – From Fundamentals to Nanoscale Dynamics* (Springer, New York, 2006).

[8]   L. Wee, R.L. Stamps, L. Malkinski, and Z. Celinski, Phys. Rev. B 69, 134426 (2004).

[9]   L. Wee, R.L. Stamps, L. Malkinski, Z. Celinski, and D. Skrzypek, Phys. Rev. B 69, 134425 (2004).

[10] O. Muller, and R. Roy, *The Major Ternary Structural Families* (Springer, NewYork-Heidelberg-Berlin, 1974).

[11] V. Scatturin, L. Corliss, N Elliott, and J. Hastings, Acta Cryst. 14, 19 (1961).

[12] K. Hirakawa, K. Hirakawa, and T. Hashimoto, J. Phys. Soc. Jpn. 15, 2063 (1960).

[13] Z. Celinski, and D. Skrzypek, Acta Phys. Pol. A 65, 149 (1984).

[14] R.A. Erickson, Phys. Rev. 90, 779 (1953).

[15] W. Stout and S.A. Reed, J. Am. Chem. Soc. 76, 5279 (1954).

[16] W. Jauch, A. Palmer and A.J. Schultz, Acta Cryst. B49, 984 (1993); K. Haefner, Ph.D. thesis, Univ. of Chicago, 1964.

[17] M.T. Hutchings, B.D.Rainford, and H.J. Guggenheim, J. Phys. C 3, 307 (1970).

[18] M.R. Fitzsimmons, B.J. Kirby. S. Roy, Zhi-Pan Li, Igor V. Roshchin, S.K. Sinha, and I.K. Schuller, Phys. Rev. B 75, 214412 (2007).

[19] J. Nogués, D. Lederman, T.J. Moran, I.K. Schuller, and K.V. Rao, Appl. Phys. Lett. 68, 3186 (1996).

[20] J. Nogués, T.J. Moran, D. Lederman, and I.K. Schuller, Phys. Rev. B 59, 6984 (1999).

[21] M.R. Fitzsimmons, P. Yashar, C. Leighton, I.K Schuller, J. Nogués, C.F. Majchrzak, and J. Dura, Phys. Rev. Lett. 84, 3986 (2000).

[22] M. R. Fitzsimmons, C. Leighton, J. Nogués, A. Hoffmann, Kai Liu, C. F. Majkrzak, J.A. Dura, J. R. Groves, R. W. Springer, P. N. Arendt, V. Leiner, H. Lauter, and I.K. Schuller, Phys. Rev. B 65, 134436 (2002).

[23] S. Widuch, Z. Celinski, K. Balin, R. Schäfer, L. Schultz, D. Skrzypek, and J. McCord, Phys. Rev. B.77, 184433 (2008).

[24] L.Malkinski, T.O'Keevan, R.E.Camley, Z.Celinski, L.Wee, R.L.Stamps, D.Skrzypek, J. Appl. Phys. 93, 6835 (2003)

[25] W.Pang, R.L.Stamps, L.Malkinski, Z.Celinski, D.Skrzypek, J. Appl. Phys. 95, 7309 (2004)

[26] N. C. Koon, Phys. Rev. Lett. 78, 4865 (1997).

[27] T. C. Schulthess, and W. H. Butler, Phys. Rev. Lett. 81, 4516 (1998); J. Appl. Phys. 85, 5510 (1999).

[28] M. D. Stiles, and R. D. McMichael, Phys Rev. B 59, 3722 (1999).
[29] M. D. Stiles, and R. D. McMichael, Phys. Rev. B 60, 12950 (1999).
[30] R. Camley, and R.L. Stamps, J. Phys.: Condens. Matter 5, 3727 (1993).

Chapter 8

Unusual Temperature Dependence of Zero-Field Ferromagnetic Resonance in Millimeter Wave Region on Al-Substituted ε-Fe₂O₃

Marie Yoshikiyo, Asuka Namai and Shin-ichi Ohkoshi

Additional information is available at the end of the chapter

1. Introduction

Insulating magnetic materials absorb electromagnetic waves. This absorption property is one of the important functions of magnetic materials, which is widely applied in our daily life as electromagnetic wave absorbers to avoid electromagnetic interference problems [1-5]. For example, spinel ferrites are used as absorbers for the present Wi-Fi communication, which uses 2.4 GHz and 5 GHz frequency waves. With the development of information technology, the demand is rising for sending heavy data such as high-resolution images at high speed. Recently, high-frequency electromagnetic waves in the frequency range of 30–300 GHz, called millimeter waves, are drawing attention as a promising carrier for the next generation wireless communication. For example, 76 GHz is an important frequency, which is beginning to be used for vehicle radars. There are also new audio products coming to use, applying millimeter wave communication in the 60 GHz region [6,7]. However, there had been no magnetic material that could absorb millimeter waves above 80 GHz before our report on ε-Fe₂O₃.

Well-known forms of Fe₂O₃ are α-Fe₂O₃ and γ-Fe₂O₃, commonly called as hematite and maghemite, respectively. However, our research group first succeeded in preparing a pure phase of ε-Fe₂O₃, which is a rare phase of iron oxide Fe₂O₃ that is scarcely found in nature [8–10]. Since then, its physical properties have been actively studied, and one of the representative properties is the gigantic coercive field (H_c) of 20 kilo-oersted (kOe) at room temperature [11–18]. We have also reported metal-substituted ε-Fe₂O₃ (ε-M_xFe₂₋ₓO₃, M = In, Ga, Al, and Rh), and showed that this series absorb millimeter waves from 35–209 GHz at room temperature due to zero-field ferromagnetic resonance (so called natural resonance) [19-29]. ε-Fe₂O₃ based magnet is expected to be a leading absorbing material for the future wireless communication using higher frequency millimeter waves.

In this chapter, we first introduce the synthesis, crystal structure, magnetic properties, and the formation mechanism of the original ε-Fe$_2$O$_3$ [8–10]. Then we report the physical properties of Al-substituted ε-Fe$_2$O$_3$, mainly focusing on its millimeter wave absorption properties due to zero-field ferromagnetic resonance. The resonance frequency was widely controlled from 112–182 GHz by changing the aluminum substitution ratio [23]. Furthermore, from a scientific point of view, temperature dependence of zero-field ferromagnetic resonance was investigated and was found to show an anomalous behavior caused by the spin reorientation phenomenon [28].

2. ε-Fe$_2$O$_3$

This section introduces the synthesis, crystal structure, magnetic properties, and the formation mechanism of ε-Fe$_2$O$_3$. ε-Fe$_2$O$_3$ had only been known as impurity in iron oxide materials, and its properties were clarified for the first time after our success in the synthesis of single-phase ε-Fe$_2$O$_3$ in 2004 [8].

2.1. Synthesis, crystal structure, and magnetic properties of ε-Fe$_2$O$_3$

Single-phase ε-Fe$_2$O$_3$ nanoparticles are synthesized by a chemical method, combining reverse-micelle and sol-gel techniques (Figure 1) [8–10,16]. In the reverse-micelle step, two reverse-micelle systems, A and B, are formed by cetyl trimethyl ammonium bromide (CTAB) and 1-butanol in n-octane. Reverse-micelle A contains aqueous solution of Fe(NO$_3$)$_3$ and Ba(NO$_3$)$_2$, and reverse-micelle B contains NH$_3$ aqueous solution. These two microemulsion systems are mixed under rapid stirring. Tetraethoxysilane (C$_2$H$_5$O)$_4$Si is then added to this solution, which forms SiO$_2$ matrix around the Fe(OH)$_3$ nanoparticles through 20 hours of stirring. The precipitation is separated by centrifugation and sintered in air at 1000°C for 4 hours. The SiO$_2$ matrix is removed by stirring in NaOH solution at 60°C for 24 hours.

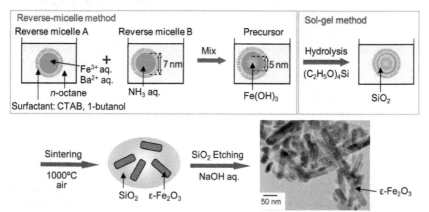

Figure 1. Synthetic procedure of ε-Fe$_2$O$_3$ nanomagnets using a combination method of reverse-micelle and sol-gel techniques. The inset is a transmission electron microscopy image of ε-Fe$_2$O$_3$ nanorods.

With this synthesis method, rod-shaped ε-Fe$_2$O$_3$ is obtained due to the effect of Ba^{2+} ions, which adsorb on particular planes of ε-Fe$_2$O$_3$, inducing growth towards one direction. Spherical ε-Fe$_2$O$_3$ nanoparticles can also be synthesized by a different method without Ba^{2+} ions, which is an impregnation method using mesoporous silica nanoparticles [17,29,30]. Methanol and water solution containing Fe(NO$_3$)$_3$ is immersed into mesoporous silica and heated in air at 1200°C for 4 hours. The etching process is the same as above.

The crystal structure of ε-Fe$_2$O$_3$ is shown in Figure 2a. It has an orthorhombic crystal structure (space group $Pna2_1$) with four non-equivalent Fe sites, A, B, C, and D sites. A, B, and C sites are six-coordinated octahedral sites, and D site is a four-coordinated tetrahedral site. ε-Fe$_2$O$_3$ exhibits spontaneous magnetization at a Curie temperature (Tc) of 500 K. Figure 2b presents magnetization versus external magnetic field curve at 300 K, which shows a huge Hc value of 20 kOe. Before this finding, the largest Hc value among metal oxide was 6 kOe of barium ferrite, BaFe$_{12}$O$_{19}$ [31], which indicates that the Hc of ε-Fe$_2$O$_3$ is over three times larger. The magnetic structure has been investigated using molecular field theory, which indicated that B and C sites have positive sublattice magnetizations, and A and D sites have negative sublattice magnetizations [32]. This result was consistent with the experimental results from neutron diffraction measurements, Mössbauer spectroscopy measurements, etc. [13,14], and was also consistent with first-principles calculation results [33].

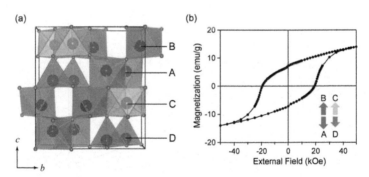

Figure 2. (a) Crystal structure of ε-Fe$_2$O$_3$. Dark blue, purple, light blue, and pink polyhedrons indicate A, B, C, and D sites, respectively. (b) Magnetization versus external magnetic field curve of ε-Fe$_2$O$_3$ at 300 K. Inset is a schematic illustration of the sublattice magnetizations of each site.

2.2. Formation mechanism of ε-Fe$_2$O$_3$

Here we discuss the formation mechanism of ε-Fe$_2$O$_3$ from the viewpoint of phase transformation. By changing the sintering temperature in the present synthesis, a phase transformation of γ-Fe$_2$O$_3$ → ε-Fe$_2$O$_3$ → α-Fe$_2$O$_3$ was observed accompanied by an increase of particle size. γ- and α-Fe$_2$O$_3$ are very common phases of Fe$_2$O$_3$, and it has been well known that γ-Fe$_2$O$_3$ transforms directly into α-Fe$_2$O$_3$ in a bulk form. In the present case, it is considered that ε-Fe$_2$O$_3$ appeared as a stable phase at an intermediate size region due to the large surface energy effect. Free energy of each i-phase (G_i, $i = \gamma$, ε, or α) is expressed as a sum of chemical potential (μ_i) and surface energy ($A_i\sigma_i$):

$$G_i = \mu_i + A_i\sigma_i, \tag{1}$$

where A_i is molar surface area and σ_i is surface free energy of a particle. Since, A_i is equal to $6V_{m,i}/d$, where $V_{m,i}$ and d represent the molar volume and particle diameter, respectively, the free energy per molar volume is expressed as

$$G_i / V_{m,i} = \mu_i / V_{m,i} + 6\sigma_i / d. \tag{2}$$

This equation indicates that the contribution of the surface energy increases with the decrease of particle diameter. When the parameters satisfy the following three conditions, $\mu_\gamma > \mu_\varepsilon > \mu_\alpha$, $\sigma_\gamma < \sigma_\varepsilon < \sigma_\alpha$, and $(\sigma_\varepsilon - \sigma_\gamma)/(\sigma_\alpha - \sigma_\varepsilon) < (\mu_\varepsilon - \mu_\gamma)/(\mu_\alpha - \mu_\varepsilon)$, the free energy curve for each phase, $G_\gamma/V_{m,\gamma}$, $G_\varepsilon/V_{m,\varepsilon}$, and $G_\alpha/V_{m,\alpha}$ intersect to form ε-Fe$_2$O$_3$ as the most stable phase at an intermediate d value (Figure 3). Such nanosize effect has also been reported for other metal oxide materials, e.g. Al$_2$O$_3$ [34,35] and Ti$_3$O$_5$ [36].

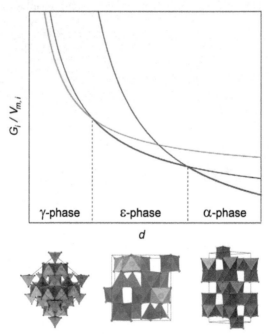

Figure 3. Representation of free energy per volume (G/V) versus particle diameter (d) for the three phases of Fe$_2$O$_3$. Green, blue, and red lines indicate the G/V curves for γ-Fe$_2$O$_3$, ε-Fe$_2$O$_3$, and α-Fe$_2$O$_3$, respectively. Below the graph are the crystal structures of γ-Fe$_2$O$_3$, ε-Fe$_2$O$_3$, and α-Fe$_2$O$_3$ from the left to right.

3. Al-substituted ε-Fe$_2$O$_3$

In this section, synthesis, crystal structure, and various physical properties of Al-substituted ε-Fe$_2$O$_3$, ε-Al$_x$Fe$_{2-x}$O$_3$, is discussed. Especially, the millimeter wave absorption property by zero-field ferromagnetic resonance is focused.

3.1. Synthesis of Al-substituted ε-Fe$_2$O$_3$

ε-Al$_x$Fe$_{2-x}$O$_3$ samples (x = 0.06, 0.09, 0.21, 0.30, 0.40) were synthesized by the same method as the original ε-Fe$_2$O$_3$, using the combination of reverse-micelle and sol-gel techniques. Reverse-micelle A contained aqueous solution of Fe(NO$_3$)$_3$ and Al(NO$_3$)$_3$, and the mixing ratio was adjusted to obtain the different samples, x = 0.06, 0.09, 0.21, 0.30, and 0.40. The sintering temperature was 1050°C for x = 0.06, 0.09, 0.30, and 0.40, and 1025°C for x = 0.21. Only the sample for x = 0 was prepared by an impregnation method using mesoporous silica nanoparticles. The SiO$_2$ matrices for all samples were etched by NaOH solution. The morphology and size of the obtained samples were examined using transmission electron microscopy (TEM), which showed spherical nanoparticles with an average particle size between 20-50 nm (Figure 4).

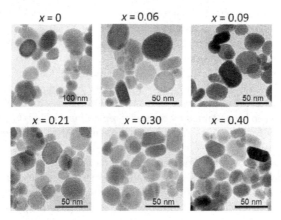

Figure 4. Transmission electron microscopy images of ε-Al$_x$Fe$_{2-x}$O$_3$ samples. The black bars indicate the scale.

3.2. Al-substitution effect in crystal structure and magnetic properties

X-ray diffraction (XRD) patterns indicated the samples to have the same orthorhombic crystal structure as the original ε-Fe$_2$O$_3$. The Rietveld analyses of the XRD patterns showed a constant decrease in the lattice constants with the degree of Al-substitution. The analysis results also indicated that the Al^{3+} ions introduced in the samples have site selectivity in the substitution. For example, in the x = 0.21 sample, the Al^{3+} substitution ratio of each Fe site was 0%, 3%, 8%, and 30% for A, B, C, and D site, respectively. This tendency for the Al^{3+} ion to prefer D site was consistent with all of the Al-substituted samples (Figure 5). This site selectivity can be understood by the smaller ion radius of Al^{3+} (0.535 Å) compared to Fe^{3+} (0.645 Å) [37]. The Al^{3+} ions prefer to occupy the smaller tetrahedral D site than the octahedral A, B, and C sites.

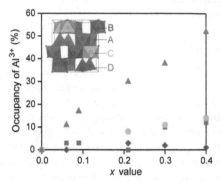

Figure 5. Al^{3+} occupancy ratio for A, B, C, and D site. Square, diamond, circle, and triangle plots represent A, B, C, and D site, respectively. Inset is the crystal structure of ε-Fe_2O_3.

The magnetic properties of the samples are shown in Table 1. The field-cooled magnetization curves under an external magnetic field of 10 Oe showed that the T_C value decreased from 500 K to 448 K with the increase of Al-substitution (Figure 6, upper right). From the magnetization versus external magnetic field measurements, gradual change of the hysteresis loops was also observed. The obtained hysteresis loops of x = 0, 0.21, and 0.40 samples are shown in Figure 6. With Al-substitution, the H_c value decreased from 22.5 kOe to 10.2 kOe, and saturation magnetization (M_s) value increased. These changes in the magnetic properties can be explained by the metal replacement of Fe^{3+} magnetic ions ($3d^5$, S = 5/2) by non-magnetic Al^{3+} ions ($3d^0$, S = 0). As mentioned previously, ε-Fe_2O_3 is a ferrimagnet with positive sublattice magnetizations at B and C sites and negative sublattice magnetizations at A and D sites. With the substitution of D site Fe^{3+} ions with non-magnetic Al^{3+}, the total magnetization increases, leading to the increase of M_s value. In addition, the non-magnetic Al^{3+} ions reduce the superexchange interaction between the magnetic sites, resulting in a decrease of T_C [32]. In this way, the magnetic properties can be widely controlled by Al-substitution.

x	T_C (K)	H_c (kOe)	M_s (emu/g)
0	500	22.5	14.9
0.06	496	19.1	15.1
0.09	490	17.5	14.6
0.21	480	14.9	17.0
0.30	466	13.8	20.3
0.40	448	10.2	19.7

Table 1. Magnetic properties of ε-$Al_xFe_{2-x}O_3$.

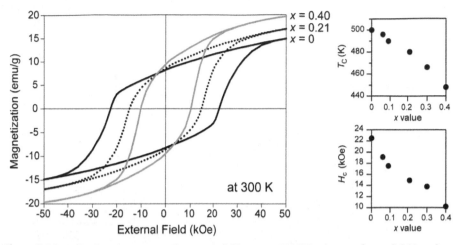

Figure 6. Magnetization versus external magnetic field curve at 300 K for the samples $x = 0$, 0.21, and 0.40 (left), Curie temperature (T_C) versus x value plot (upper right), and coercive field (H_c) versus x value plot (lower right).

3.3. Electromagnetic wave absorption of Al-substituted ε-Fe₂O₃ by zero-field ferromagnetic resonance

Zero-field ferromagnetic resonance is a resonance phenomenon caused by the gyromagnetic effect induced by an electromagnetic wave irradiation under no magnetic field (Figure 7). This phenomenon is observed in ferromagnetic materials with magnetic anisotropy. When the magnetization is tilted away from the easy-axis by the magnetic component of the electromagnetic wave, precession of the magnetization occurs around the easy-axis due to gyromagnetic effect. Resonance is observed when this precession frequency coincides with the electromagnetic wave frequency, resulting in electromagnetic wave absorption at the particular frequency [38]. This resonance frequency (f_r) is proportional to the magnetocrystalline anisotropy (H_a) and can be expressed as

$$f_r = (v / 2\pi)H_a, \tag{3}$$

where v is the gyromagnetic ratio. If the sample is consisted of randomly oriented particles with uniaxial magnetic anisotropy, the H_a value is proportional to H_c. Therefore, electromagnetic wave absorption at high frequencies is expected with insulating materials exhibiting large coercivity, which is the case for ε-Fe₂O₃ based magnets.

With the general electromagnetic wave absorption measurement using free space absorption measurement system, the absorption frequencies of the present ε-Al$_x$Fe$_{2-x}$O₃ samples exceeded the measurement range, where the maximum is 110 GHz. Therefore, the absorption measurements were conducted using terahertz time domain spectroscopy (THz-TDS) at room temperature. The THz-TDS measurement system is shown in Figure 8. A

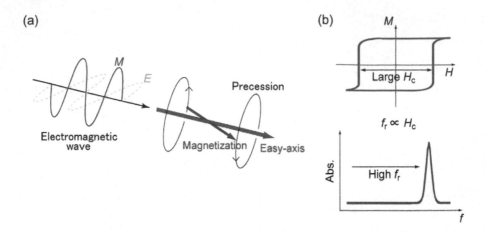

Figure 7. (a) A schematic illustration of zero-field ferromagnetic resonance (natural resonance), resulting in electromagnetic wave absorption due to the precession of magnetization around the easy-axis. The M and E of the electromagnetic wave indicate the magnetic and electric components, respectively. (b) An illustration indicating that larger coercive field (H_c) results in higher resonance frequency (f_r).

mode-locked Ti:sapphire femtosecond pulse laser with a time duration of 20 fs at a repetition rate of 76 MHz was used. The output was divided into a pump and probe beam for the time-domain system. For THz wave emitter and detector, dipole type and bowtie type low-temperature-grown GaAs photoconductive antennas were used, respectively. The sample was set on a sample holder, which was inserted between a set of paraboloidal mirrors concentrating the THz wave at the location of the sample. The temporal waveforms of the electric component of the transmitted THz pulse waves were obtained by changing the delay time between the pump and probe pulses. The temporal waves were Fourier transferred to obtain the frequency dependence, and the absorption spectra were calculated using the following equation:

$$(Absorption) = -10\log|t(\omega)|^2 \text{ (dB)}, \tag{4}$$

where $t(\omega)$ is the complex amplitude transmittance. An absorption of 20 dB indicates 99% absorption.

The electromagnetic wave absorption spectra are shown in Figure 9. Absorption peaks were observed at 112 GHz ($x = 0.40$), 125 GHz ($x = 0.30$), 145 GHz ($x = 0.21$), 162 GHz ($x = 0.09$), 172 GHz ($x = 0.06$), and 182 GHz ($x = 0$). The f_r value decreased with Al-substitution, consistent with the behavior of the H_c value (Figure 6, lower right). The observed electromagnetic wave absorption due to zero-field ferromagnetic resonance at exceptional high frequencies was achieved by the large H_a value of this series with large coercivity.

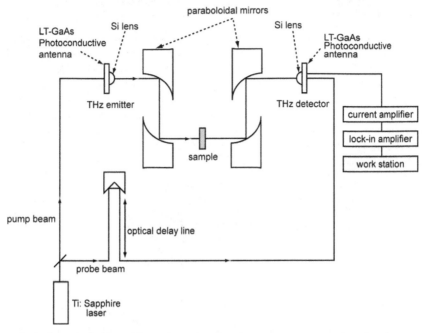

Figure 8. A schematic diagram of the terahertz time domain spectroscopy measurement system.

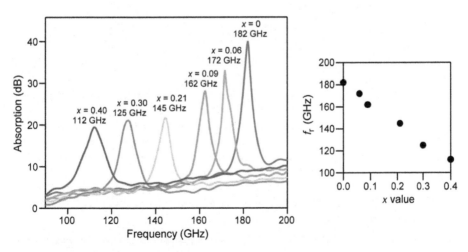

Figure 9. Electromagnetic wave absorption spectra of ε-Al$_x$Fe$_{2-x}$O$_3$ (left). Red, orange, yellow, green, light blue, and blue lines are the absorption spectra for $x = 0.40, 0.30, 0.21, 0.09, 0.06$, and 0, respectively. Right is the zero-field ferromagnetic resonance frequency (f_r) versus x value plot.

3.4. Temperature dependence of zero-field ferromagnetic resonance in Al-substituted ε-Fe$_2$O$_3$

Among the ε-Al$_x$Fe$_{2-x}$O$_3$ samples discussed in the previous section, we focused on the x = 0.06 sample and measured the temperature dependence of zero-field ferromagnetic resonance. The Al^{3+} substitution ratios of each Fe site in ε-Al$_{0.06}$Fe$_{1.94}$O$_3$ are 3%, 0%, 0%, and 11% for A, B, C, and D site, respectively. The magnetic properties of ε-Al$_{0.06}$Fe$_{1.94}$O$_3$ are shown in Figure 10. The field-cooled magnetization curve under 10 Oe external magnetic field showed a T_C value of 496 K and a cusp at 131 K (= T_P). The cusp in the magnetization is due to the spin reorientation phenomenon, which is known to occur in this temperature region [11,12]. The magnetization versus external magnetic field curve exhibited an H_c value of 19.1 kOe at 300 K.

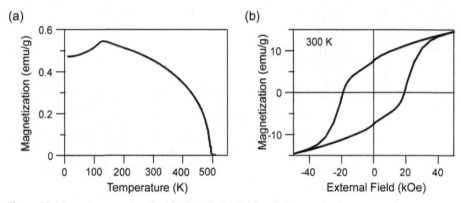

Figure 10. Magnetic properties of ε-Al$_{0.06}$Fe$_{1.94}$O$_3$. (a) Field-cooled magnetization curve under an external magnetic field of 10 Oe. (b) Magnetization versus external magnetic field curve at 300 K.

For the THz-TDS measurement, ε-Al$_{0.06}$Fe$_{1.94}$O$_3$ powder sample was pressed into a pellet-form. The absorption spectra at different temperatures are shown in Figure 11a. These absorption spectra versus frequency were obtained by calibration of the background noise. They were also fitted by Lorentz function. At 301 K, the f_r value was 172 GHz, consistent with the result in the previous section. With the decrease of temperature, the f_r value gradually increased to 186 GHz at 204 K, and turned to an abrupt decrease down to 147 GHz at 77 K. The f_r value continued to decrease with lowering the temperature, and at 21 K, the f_r value was 133 GHz (Figure 11b). Temperature dependence was also observed in the linewidth of the absorption spectra. The full width at half maximum (Δf) value increased from 5 GHz at 301 K to 19 GHz at 77 K with decreasing temperature, and then, decreased to 16 GHz at 21 K (Figure 11c).

Temperature dependencies of magnetic hysteresis loop and ac magnetic susceptibility was studied in order to understand the anomalous temperature dependencies of f_r and Δf. As mentioned in Figure 10a, the field-cooled magnetization curve shows an increase below T_C, but a cusp appears at T_P = 131 K, where the magnetization turns to a decrease. The H_c value increased from 19.4 kOe at 300 K to 22.6 kOe at 200 K, and then decreased to 4.5 kOe at 70 K with the decrease of temperature. The H_c versus temperature plot indicates a sigmoid decrease in a wide temperature range of 200 – 60 K centered at T_P (i.e., ±70 K from the center temperature, T_P = 131 K) (Figure 12a). In other words, the beginning and ending temperatures of the spin reorientation are about 200 K and 60 K, respectively, with decreasing temperature. The temperature region of the sigmoid decrease of f_r almost corresponds to the temperature range of the spin reorientation. The sigmoid increase of Δf

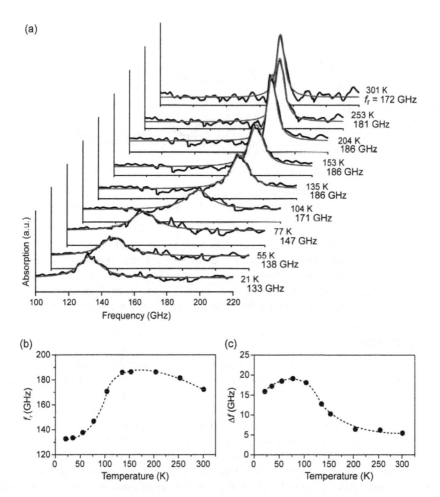

Figure 11. (a) Electromagnetic wave absorption spectra of ε-Al$_{0.06}$Fe$_{1.94}$O$_3$ at different temperatures. Black lines and red lines indicate the observed spectra and fitted Lorentz function. (b) Temperature dependence of zero-field ferromagnetic resonance frequency (f_r). (c) Temperature dependence of full width at half maximum (Δf) of the absorption spectra. Dotted lines are to guide the eye.

was also observed in the spin reorientation temperature region. Figure 12b is the ac magnetic susceptibility versus temperature with frequency of 10 Hz under field amplitude of 1 Oe. As the temperature decreased, the real part of the ac magnetic susceptibility (χ') gradually increased to a maximum value of 3.8×10^{-4} emu/g·Oe at 60 K and then decreased. The imaginary part (χ'') showed similar temperature dependence with a maximum around 70 K. These temperature dependencies of ac magnetic susceptibility correspond to that of Δf [39,40].

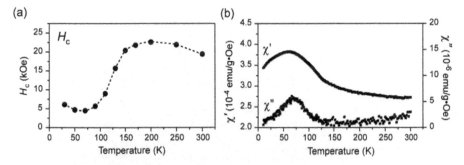

Figure 12. (a) Temperature dependence of coercive field (H_c). Dotted line is to guide the eye. (b) Temperature dependence of ac magnetic susceptibility (real part χ' and imaginary part χ'') measured at 10 Hz and 1 Oe field amplitude.

As mentioned previously, the f_r value is proportional to the H_a value, and in this case with randomly oriented samples, f_r is also related to the H_c value. Therefore, the observed anomalous temperature dependence of f_r in ε-Al$_{0.06}$Fe$_{1.94}$O$_3$ was understood by the temperature dependence of H_c. The sigmoid decrease centered at T_p originates from the disappearance of magnetic anisotropy due to the spin reorientation phenomenon [11–13].

4. Conclusion

In this chapter, a rare phase of diiron trioxide, ε-Fe$_2$O$_3$, and its Al-substituted series were introduced. The synthesis, crystal structure, and its exceptional physical properties were discussed, especially its huge magnetic anisotropy exhibiting a gigantic coercive field, which enables electromagnetic wave absorption due to zero-field ferromagnetic resonance at high frequencies in the millimeter wave region. Al-substitution effect was observed in the ε-Al$_x$Fe$_{2-x}$O$_3$ series, widely controlling the magnetic properties and the zero-field ferromagnetic resonance frequency: ε-Al$_x$Fe$_{2-x}$O$_3$ absorbed millimeter waves from 112–182 GHz at room temperature. Temperature dependence of zero-field ferromagnetic resonance was also investigated for ε-Al$_{0.06}$Fe$_{1.94}$O$_3$ sample, and an anomalous behavior was observed due to spin reorientation phenomenon.

Since ε-Al$_x$Fe$_{2-x}$O$_3$ is composed of very common and low costing elements, it is friendly to the environment and can be economically produced. Its chemical stability is also an advantage in the viewpoint of industrial applications, such as electromagnetic wave absorbers in the near future, where high-frequency millimeter waves are likely to be used in order to transport heavy data at high speed.

Author details

Marie Yoshikiyo, Asuka Namai and Shin-ichi Ohkoshi
Department of Chemistry, School of Science, The University of Tokyo, Tokyo, Japan

Shin-ichi Ohkoshi
CREST, JST, K's Gobancho, 7 Gobancho, Chiyoda-ku, Tokyo, Japan

Acknowledgement

The present research was supported partly by the Core Research for Evolutional Science and Technology (CREST) program of the Japan Science and Technology Agency (JST), a Grant-in-Aid for Young Scientists (S) from Japan Society for the Promotion of Science (JSPS), DOWA Technofund, the Asahi Glass Foundation, Funding Program for Next Generation World-Leading Researchers from JSPS, a Grant for the Global COE Program "Chemistry Innovation through Cooperation of Science and Engineering", Advanced Photon Science Alliance (APSA) from the Ministry of Education, Culture, Sports, Science and Technology of Japan (MEXT), the Cryogenic Research Center, The University of Tokyo, and the Center for Nano Lithography & Analysis, The University of Tokyo, supported by MEXT Japan. M. Y. is grateful to Advanced Leading Graduate Course for Photon Science (ALPS) and JSPS Research Fellowships for Young Scientists. A. N. is grateful to JSPS KAKENHI Grant Number 24850004 and Office for Gender Equality, The University of Tokyo. We are grateful to Dr. S. Sakurai of The University of Tokyo. We also thank Prof. M. Nakajima and Prof. T. Suemoto for support in THz-TDS measurements, Mr. Y. Kakegawa and Mr. H. Tsunakawa for collecting the TEM images, and Mr. K. Matsumoto, Mr. M. Goto, Mr. S. Sasaki, Mr. T. Miyazaki, and Mr. T. Yoshida of DOWA Electronics Materials Co., Ltd. for the valuable discussions.

5. References

[1] Zhou ZG. Magnetic Ferrite Materials. Beijing: Science Press; 1981.

[2] Yoshida S, Sato M, Sugawara E, Shimada Y. Permeability and Electromagnetic-Interference Characteristics of Fe-Si-Al Alloy Flakes-Polymer Composite. J. Appl. Phys. 1999;85(8) 4636–4638.

[3] Naito Y, Suetake K. Application of Ferrite to Electromagnetic Wave Absorber and its Characteristics. IEEE Trans. Microwave Theory Tech. 1971;MT19(1) 65–72.

[4] Yusoff AN, Abdullah MH, Ahmed SH, Jusoh SF, Mansor AA, Hamid SAA. Electromagnetic and Absorption Properties of Some Microwave Absorbers. J. Appl. Phys. 2002;92 (2) 876–882.

[5]   Adam JD, Davis LE, Dionne GF, Schloemann EF, Stitzer SN. Ferrite Devices and Materials. IEEE Trans. Microwave Theory Tech. 2002;50(3) 721–737.

[6]   Federici J, Moeller L. Review of Terahertz and Subterahertz Wireless Communications. J. Appl. Phys. 2010;107(11) 111101.

[7]   Khatib M. Wireless Communications. Rijeka: Intech; 2011.

[8]   Jin J, Ohkoshi S, Hashimoto K. Giant Coercive Field of Nanometer-Sized Iron Oxide. Adv. Mater. 2004;16(1) 48–51.

[9]   Ohkoshi S, Sakurai S, Jin J, Hashimoto K. The Addition Effects of Alkaline Earth Ions in the Chemical Synthesis of ε-Fe$_2$O$_3$ Nanocrystals that Exhibit a Huge Coercive Field. J. Appl. Phys. 2005;97(10) 10K312.

[10] Jin J, Hashimoto K, Ohkoshi S. Formation of Spherical and Rod-Shaped ε-Fe$_2$O$_3$ Nanocrystals with a Large Coercive Field. J. Mater. Chem. 2005;15(19) 1067–1071.

[11] Sakurai S, Jin J, Hashimoto K, Ohkoshi S. Reorientation Phenomenon in a Magnetic Phase of ε-Fe$_2$O$_3$ Nanocrystal. J. Phys. Soc. Jpn. 2005;74(7) 1946–1949.

[12] Kurmoo M, Rehspringer J, Hutlova A, D'Orleans C, Vilminot S, Estournes C, Niznansky D. Formation of Nanoparticles of ε-Fe$_2$O$_3$ from Yttrium Iron Garnet in a Silica Matrix: An Unusually Hard Magnet with a Morin-Like Transition below 150 K. Chem. Mater. 2005;17(5) 1106–1114.

[13] Gich M, Frontera C, Roig A, Taboada E, Molins E, Rechenberg HR, Ardisson JD, Macedo WAA, Ritter C, Hardy V, Sort J, Skumryev V, Nogués J. High- and Low-Temperature Crystal and Magnetic Structures of ε-Fe$_2$O$_3$ and their Correlation to its Magnetic Properties. Chem. Mater. 2006;18(16) 2889–2897.

[14] Tucek J, Ohkoshi S, Zboril R. Room-Temperature Ground Magnetic State of ε-Fe$_2$O$_3$: In-Field Mössbauer Spectroscorpy Evidence for Collinear Ferrimagnet. Appl. Phys. Lett. 2011;99(25) 253108.

[15] Sakurai S, Shimoyama J, Hashimoto K, Ohkoshi S. Large Coercive Field in Magnetic-Field Oriented ε-Fe$_2$O$_3$ Nanorods. Chem. Phys. Lett. 2008;458(4–6) 333–336.

[16] Sakurai S, Tomita K, Hashimoto K, Yashiro H, Ohkoshi S. Preparation of the Nanowire Form of ε-Fe$_2$O$_3$ Single Crystal and a Study of the Formation Process. J. Phys. Chem. C 2008;112(34) 13095–13098.

[17] Sakurai S, Namai A, Hashimoto K, Ohkoshi S. First Observation of Phase Transformation of All Four Fe$_2$O$_3$ Phases ($\gamma \rightarrow \varepsilon \rightarrow \beta \rightarrow \alpha$-Phase). J. Am. Chem. Soc. 2009;131(51) 18299–18303.

[18] Tucek J, Zboril R, Namai A, Ohkoshi S. ε-Fe$_2$O$_3$: An Advanced Nanomaterial Exhibiting Giant Coercive Field, Millimeter-Wave Ferromagnetic Resonance, and Magnetoelectric Coupling. Chem. Mater. 2010;22(24) 6483–6505.

[19] Sakurai S, Kuroki S, Tokoro H, Hashimoto K, Ohkoshi S. Synthesis, Crystal Structure, and Magnetic Properties of ε-In$_x$Fe$_{2-x}$O$_3$ Nanorod-Shaped Magnets. Adv. Funct. Mater. 2007;17(14) 2278–2282.

[20] Namai A, Sakurai S, Ohkoshi S. Synthesis, Crystal Structure, and Magnetic Properties of ε-Ga$^{III}_x$Fe$^{III}_{2-x}$O$_3$ Nanorods. J. Appl. Phys. 2009;105(7) 07B516.

[21] Yamada K, Tokoro H, Yoshikiyo M, Yorinaga T, Namai A, Ohkoshi S. Phase Transition of ε-In$_x$Fe$_{2-x}$O$_3$ Nanomagnets with a Large Thermal Hysteresis Loop. J. Appl. Phys. 2012;111(7) 07B506.

[22] Ohkoshi S, Kuroki S, Sakurai S, Matsumoto K, Sato K, Sasaki S. A Millimeter-Wave Absorber Based on Gallium-Substituted ε-Iron Oxide Nanomagnets. Angew. Chem. Int. Ed. 2007;46(44) 8392–8395.

[23] Namai A, Sakurai S, Nakajima M, Suemoto T, Matsumoto K, Goto M, Sasaki S, Ohkoshi S. Synthesis of an Electromagnetic Wave Absorber for High-Speed Wireless Communication. J. Am. Chem. Soc. 2009;131(3) 1170–1173.

[24] Namai A, Kurahashi S, Hachiya H, Tomita K, Sakurai S, Matsumoto K, Goto T, Ohkoshi S. High Magnetic Permeability of ε-Ga$_x$Fe$_{2-x}$O$_3$ Magnets in the Millimeter Wave Region. J. Appl. Phys. 2010;107(9) 09A955.

[25] Nakajima M, Namai A, Ohkoshi S, Suemoto T. Ultrafast Time Domain Demonstration of Bulk Magnetization Precession at Zero Magnetic Field Ferromagnetic Resonanace Induced by Terahertz Magntic Field. Opt. Express 2010;18(17) 18260–18268.

[26] Afsar MN, Korolev KA, Namai A, Ohkoshi S. Measurements of Complex Magnetic Permeability of Nano-Size ε-Al$_x$Fe$_{2-x}$O$_3$ Powder Materials at Microwave and Millimeter Wavelengths. IEEE Trans. Magn. 2012;48(11) 2769–2742.

[27] Namai A, Kurahashi S, Goto T, Ohkoshi S. Theoretical Design of a High-Frequency Millimeter Wave Absorbing Sheet Composed of Gallium Substituted ε-Fe$_2$O$_3$ Nanomagnet. IEEE Trans. Magn. 2012;48(11) 4386–4389.

[28] Yoshikiyo M, Namai A, Nakajima M, Suemoto T, Ohkoshi S. Anomalous Behavior of High-Frequency Zero-Field Ferromagnetic Resonance in Aluminum-Substituted ε-Fe$_2$O$_3$. J. Appl. Phys. 2012;111(7) 07A726.

[29] Namai A, Yoshikiyo M, Yamada K, Sakurai S, Goto T, Yoshida T, Miyazaki T, Nakajima M, Suemoto T, Tokoro H, Ohkoshi S. Hard Magnetic Ferrite with a Gigantic Coercivity and High Frequency Millimetre Wave Rotation. Nat. Commun. 2012;3 1035.

[30] Möller K, Kobler J, Bein T. Collidal Suspensions of Nanometer-Sized Mesoporous Silica. Adv. Funct. Mater. 2007;17(4) 605–612.

[31] Haneda K, Miyakawa C, Kojima H. Preparation of High-Coercivity BaFe$_{12}$O$_{19}$. J. Am. Ceram. Soc. 1974;57(8) 354–357.

[32] Ohkoshi S, Namai A, Sakurai S. The Origin of Ferromagnetism in ε-Fe$_2$O$_3$ and ε-Ga$_x$Fe$_{2-x}$O$_3$ Nanomagnets. J. Phys. Chem. C 2009;113(26) 11235–11238.

[33] Yoshikiyo M, Yamada K, Namai A, Ohkoshi S. Study of the Electronic Structure and Magnetic Properties of ε-Fe$_2$O$_3$ by First-Principles Calculation and Molecular Orbital Calculations. J. Phys. Chem. C 2012;116(15) 8688–8691.

[34] McHale JM, Auroux A, Perrotta AJ, Navrotsky A. Surface Energies and Thermodynamic Phase Stability in Nanocrystalline Aluminas. Science 1997;277(5327) 788–791.

[35] Wen HL, Chen YY, Yen FS, Huang CH. Size Characterization of θ- and α-Al$_2$O$_3$ Crystallites During Phase Transformation. Nanostruct. Mater. 1999;11(1) 89–101.

[36] Ohkoshi S, Tsunobuchi Y, Matsuda T, Hashimoto K, Namai A, Hakoe F, Tokoro H. Synthesis of a Metal Oxide with a Room-Temperature Photoreversible Phase Transition. Nat. Chem. 2010;2(7) 539–545.

[37] Shannon RD. Revised Effective Ionic Radii and Systematic Studies of Interatomic Distances in Halides and Chalcogenides. Acta Cryst. 1976;A32(Sep1) 751–767.

[38] Chikazumi S. Physics of Ferromagnetism. New York: Oxford University Press; 1997.

[39] Algarabel PA, Moral A, Ibarra MR, Arnaudas JI. Spin Reorientation in $ReCo_5$ Compounds: A.C. Susceptibility and Thermal Expansion. J. Phys. Chem. Solids 1988;49(2) 213–222.

[40] Malik SK, Adroja DT, Ma BM, Boltich EB, Sohn JG, Sankar SG, Wallace WE. Spin Reorientation Phenomenon in $Nd_{0.5}Er_{1.5}Fe_{14-x}M_xB$ (M = Al and Co), as Determined by AC Susceptibility Measurements. J. Appl. Phys. 1990;67(9) 4589–4591.

Optical Properties of Antiferromagnetic/Ion-Crystalic Photonic Crystals

Shu-Fang Fu and Xuan-Zhang Wang

Additional information is available at the end of the chapter

1. Introduction

Increasing attention has been paid to magnetic photonic crystals (MPCs) because the properties of the MPCs can be modulated not only with the change of their structure (including components, layer thickness or thickness ratio) but also with the external magnetic field. MPCs are capable of acting as tunable filters [1] at different frequencies, and that controllable gigantic Faraday rotation angles [2-6] are simultaneously obtained. The nonmagnetic media in MPCs generally are ordinary dielectrics, so the electromagnetic wave modes are just magnetic polaritons. The effect of magnetic permeability and dielectric permittivity of two component materials in MPCs on the photonic band groups were discussed, where the permeability and permittivity were considered as scalar quantities [7].

Recently, our group investigated the optical properties of antiferromagnetic/ ion-crystal (AF/IC) PCs [8-11]. It is well known that the two resonant frequencies of AFs, such as, FeF2 and MnF2, fall into the millimeter or far infrared frequencies regions and some ionic semiconductors possess a very low phonon-resonant frequency range like the AFs. Especially, these frequency regions also are situated the working frequency range of THz technology, so the AF/IC PCs may be available to make the new elements in the field of THz technology. Note that in ICs, including ionic semiconductors, when the frequencies of the phonon and the transverse optical (TO) phonon modes of ICs are close, the dispersion curves of phonon and TO phonon modes will be changed and a kind of coupled mode called phonon polariton will be formed. Therefore, in the AF/IC PCs, the TO phonon modes of ICs can directly couple with the electric field in an electromagnetic wave and this coupling generates the phonon polaritons, however, the magnetization's motion in magnets can directly couple with the magnetic field, which is the origin of magnetic polaritons. Thus in such an AF/IC PCs, we refer to collective polaritons as the magneto-phonon polaritons (MPPs). In the presence of external magnetic field and damping, MPPs spectra display two petty bulk mode bands with negative group velocity. It is worthy of mentioning that many surface modes emerge in the vicinity of two petty bulk mode bands, and that some surface modes bear nonreciprocality [11]. The optical properties of the AF/IC PCs can be modulated by an external magnetic field.

In addition, we have concluded that there is a material match of an AF and an IC, for which a common frequency range is found, in which the AF has a negative magnetic permeability and the IC has negative dielectric permittivity [10]. Consequently, the AF/IC structures are thought to be of the left-handed materials (LHMs) which have attracted much attention from the research community in recent years because of their completely different properties from right-handed materials (RHMs). In a LHM, the electric field, magnetic field and wave vector of a plane electromagnetic wave form a left-handed triplet, the energy flow of the plane wave is opposite in direction to that of the wave vector [12-17]. LHMs have to be constructed artificially since there is no natural LHM. Several variations of the design have been studied through experiments [18-20]. Up to now, scientists have found some LHMs available in infrared and visible ranges [21-25], but each design has a rather complicated structure. We noticed a work that discussed the left-handed properties of a superlattice composed of alternately semiconductor and antiferromagnetic (AF) layers, where the interaction between AF polaritons and semiconductor plasmons lead to the left-handedness of the superlattice [26]. However the plasmon resonant frequency sensitively depends on the free charge carrier's density, or impurity concentration in semiconductor layers, so if one wants to see a plasmon resonant frequency near to AF resonant frequencies, the density must be very low since AF resonant frequencies are distributed in the millimeter to far infrared range. In the case of such a low density, the effect of the charge carriers on the electromagnetic properties may be very weak [27] so that there is not the left-handedness of the superlattice. According the discussion above, we propose a simple structure of multilayer which consists of AF and IC layers. An analytical condition under which both left-handeness and negative refraction phenomenon appear in the film is established by calculating the angle between the energy flow and wave vector of a plane electromagnetic wave in AF/IC PCs and its refraction angle.

2. Magneto-phonon polaritons (MPPs) in AF/IC PCs

Polaritons in solids are a kind of electromagnetic modes determining optical or electromagnetic properties of the solids. Natures of various polaritons, including the surface and bulk polaritons, were very clearly discussed in Ref. [28]. Recent years, based on magnetic multilayers or superlattices, where nonmagnetic layers are of ordinary dielectric and their dielectric function is a constant, the polaritons in these structures called the MPCs were discussed [29-34]. On the other hand, ones were interested in the phonon polaritons [35-36], where the surface polariton modes could be focused by a simple way and probably possess new applications. In this part, the collective polaritons, MPPs in a superlattice structure comprised of alternating AF and IC layers, will be discussed. In the past, for simplicity, the damping was generally ignored in the discussion of dispersion properties regarding the polaritons[30,32-34]. Actually, most materials are dispersive and absorbing. Therefore, it is also necessary to consider the effect of damping.

2.1. MPPs in one-dimension AF/IC PCs

An interesting configuration in experiment is the Voigt geometry as illustrated in Fig.1, where the polariton wave propagates in the x-y plane and the magnetic field of an electromagnetic wave is parallel to this plane, but the wave electric field aligns the z direction. We concentrate our attention on the case where the external magnetic field and AF anisotropy axis both are along the z axis and parallel to layers. The y axis is perpendicular to layers in the structure. The semi-space ($y < 0$) is of vacuum, where d_a and d_i are thicknesses of AF and IC layers, respectively. For the far infrared wave, the order of the wavelength is about $100 \mu m$. Thus, as long as the thicknesses of AF and IC layers are less than $10 \mu m$, the wavelength λ will much longer than the period of AF/IC PCs. With this condition, the AF/IC PCs will become a uniform film by means of an effective-medium method (EMM).

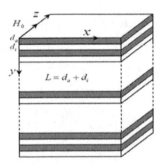

Figure 1. Illustration and coordinate system.

2.1.1. EMM for one-dimensional AF/IC PCs

We first present the permeability of the AF film. In the external magnetic field $\vec{H}_0$, the magnetic permeability is well-known, with its nonzero elements [33, 37]

$$\mu_{xx} = \mu_{yy} = \mu = 1 + \omega_m \omega_a \{[\omega_r^2 - (\omega_0 - \omega - i\tau)^2]^{-1} + [\omega_r^2 - (\omega_0 + \omega + i\tau)^2]^{-1}\}, \tag{1}$$

$$\mu_{xy} = -\mu_{yx} = \mu_{\perp} = i\omega_m \omega_a \{[\omega_r^2 - (\omega_0 - \omega - i\tau)^2]^{-1} - [\omega_r^2 - (\omega_0 + \omega + i\tau)^2]^{-1}\}. \tag{2}$$

with $\omega_0 = \gamma H_0$, $\omega_m = 4\pi\gamma M_0$, $\omega_a = \gamma H_a$, $\omega_e = \gamma H_e$, and $\omega_r = [\omega_a(2\omega_e + \omega_a)]^{1/2}$, where M_0 is the sublattice magnetization, H_a represents the anisotropy field, and H_e the exchange field. ω_r is the AF resonant frequency, γ the gyromagnetic ratio, and τ the magnetic damping constant. We use ε_a as the dielectric constant of the AF. Subsequently, we present the dielectric function of the IC [38],

$$\varepsilon_i = \varepsilon_h + \frac{(\varepsilon_l - \varepsilon_h)\omega_T^2}{\omega_T^2 - \omega^2 - i\eta\omega}, \tag{3}$$

Where ε_h and ε_l are the high- and low-frequency dielectric constants, but ω_T is the TO resonant frequency of $k=0$ and η is the phonon damping coefficient. The IC is nonmagnetic, so its magnetic permeability is taken as $\mu_i = 1$.

We assume that there are an effective relation $\vec{B} = \ddot{\mu}_{eff} \cdot \vec{H}$ between effective magnetic induction and magnetic field, and an effective relation $\vec{D} = \ddot{\varepsilon}_{eff} \cdot \vec{E}$ between effective electric field and displacement, where these fields are considered as the wave fields in the structures. But $\vec{b} = \ddot{\mu} \cdot \vec{h}$ and $\vec{d} = \varepsilon\vec{e}$ in any layer, where $\ddot{\mu}$ is given in Eqs.(1) for AF layer and $\ddot{\mu} = 1$ for IC layers. These fields are local fields in the layers. For the components of magnetic induction and field continuous at the interface, one assumes

$$H_x = h_{1x} = h_{2x}, H_z = h_{1z} = h_{2z}, B_y = b_{1y} = b_{2y}, \tag{4}$$

And for those components discontinuous at the interface, one assumes

$$B_x = f_a b_{1x} + f_i b_{2x}, B_z = f_a b_{1z} + f_i b_{2z}, H_y = f_a h_{1y} + f_i h_{2y}. \tag{5}$$

where the AF volume fraction $f_a = d_a / L$ and the IC volume fraction $f_i = d_i / L$ with the period $L = d_a + d_i$. Thus the effective magnetic permeability is achieved from Eqs. (4),(5) and it is definite by $\vec{B} = \ddot{\mu}_{eff} \cdot \vec{H}$,

$$\ddot{\mu}_{eff} = \begin{pmatrix} \mu_{xx}^e & i\mu_{xy}^e & 0 \\ -i\mu_{xy}^e & \mu_{yy}^e & 0 \\ 0 & 0 & 1 \end{pmatrix}, \tag{6}$$

with the elements

$$\mu_{xx}^e = f_a\mu + f_i - (f_a f_i \mu_\perp^2)/(\mu f_i + f_a), \mu_{yy}^e = \mu/(\mu f_i + f_a), \mu_{xy}^e = \mu_{yx}^e = f_a\mu_\perp/(\mu f_i + f_a), \tag{7}$$

On the similar principle, we can find that the effective dielectric permittivity tensor is diagonal and its elements are

$$\varepsilon_{xx}^e = \varepsilon_{zz}^e = f_a\varepsilon_a + f_i\varepsilon_i, \varepsilon_{yy}^e = \varepsilon_a\varepsilon_i / (f_a\varepsilon_a + f_i\varepsilon_i). \tag{8}$$

On the base of these effective permeability and permittivity, one can consider the AF/IC PCs as homogeneous and anisotropical AF films or bulk media. The similar discussions can be found in the Chapter 3 of the book "Propagation of Electromagnetic Waves in Complex Matter" edited by Ahmed Kishk [39].

2.1.2. Dispersion relations of surface and bulk MPP with transfer matrix method (TMM)

The wave electric fields in an AF layer and IC layer are written as

$$\vec{E} = (E_A e^{\alpha y} + E_B e^{-\alpha y})\exp(ikx - i\omega t)\vec{e}_z, \tag{9}$$

$$\vec{E} = (E_C e^{\beta y} + E_D e^{-\beta y})\exp(ikx - i\omega t)\vec{e}_z, \tag{10}$$

respectively. k is the wave-vector component along x axis. α and β are the decay coefficients when they are real, otherwise they correspond to the y wave-vector components. E_j (j=A, B, C or D) denotes the amplitudes of the electric fields. Additionally, the corresponding magneticfields can be found with the relation $\nabla \times \vec{E} = i\omega\vec{B}$. The wave equation resulting from the Maxwell equations is

$$\nabla(\nabla \cdot \vec{E}) - \nabla^2 \vec{E} - \mu_0 \omega^2/c^2 \, \vec{\varepsilon} \cdot \vec{E} = 0. \tag{11}$$

We see from the wave equation that

$$\alpha^2 = k^2 - \varepsilon_a \mu_v \omega^2/c^2 \tag{12}$$

with $\mu_v = (\mu^2 - \mu_\perp^2)/\mu$ and c is the light velocity in vacuum, and

$$\beta^2 = k^2 - \varepsilon_i \omega^2/c^2. \tag{13}$$

Employing the well-known TMM, together withthe boundary conditions of E_z and H_x continuous at the interfaces, we can find a matrix relation between wave amplitudes in any two adjacent bi-layers, or the relation between amplitudes in the nth and n+1th bi-layers

$$\begin{pmatrix} E_{an+1} \\ E_{bn+1} \end{pmatrix} = T \begin{pmatrix} E_{an} \\ E_{bn} \end{pmatrix} \tag{14}$$

where T is the transfer matrix expected and its components are

$$T_{11} = \frac{e^{\alpha d_a}}{\Delta - \Delta'}[(\Delta - \Delta')\cosh(\beta d_i) + (1 - \Delta\Delta')\sinh(\beta d_i)], \quad T_{12} = \frac{e^{-\alpha d_a}(1 - \Delta'^2)}{\Delta - \Delta'}\sinh(\beta d_i), \tag{15}$$

$$T_{21} = \frac{e^{\alpha d_a}(\Delta^2 - 1)}{\Delta - \Delta'}\sinh(\beta d_i), \quad T_{22} = \frac{e^{-\alpha d_a}}{\Delta - \Delta'}[(\Delta - \Delta')\cosh(\beta d_i) + (\Delta\Delta' - 1)\sinh(\beta d_i)], \tag{16}$$

with $\Delta = (\alpha\mu - k\mu_\perp)/[\beta(\mu^2 - \mu_\perp^2)]$ and $\Delta' = -(\alpha\mu + k\mu_\perp)/[\beta(\mu^2 - \mu_\perp^2)]$. The Bloch's theorem implies another relation

$$\begin{pmatrix} E_{an+1} \\ E_{bn+1} \end{pmatrix} = e^{iQL} \begin{pmatrix} E_{an} \\ E_{bn} \end{pmatrix}. \tag{17}$$

Based on matrix relations (14) and (17), we obtain the polariton dispersion equation

$$\cos(QL) = \cosh(\alpha d_a)\cosh(\beta d_i) + (\frac{\beta\mu_v}{2\alpha} + \frac{k^2 - \varepsilon_a\mu\omega^2/c^2}{2\alpha\beta\mu})\sinh(\alpha d_a)\sinh(\beta d_i). \qquad (18)$$

Q is the Bloch wave number for an infinite structure and is a real number, and then equation (18) describes the bulk polariton modes.

For a semi-infinite structure, it is interesting in physics that $Q = i\rho$ is an imaginary number. Thus equation (18) can be used to determine the surface modes traveling along the x axis. We need the electromagnetic boundary conditions at the surface of this structure to find another necessary equation for the surface polariton modes. This equation is just

$$T_{11} + \varphi T_{12} = \exp(-\rho L) = \varphi^{-1}T_{21} + T_{22}, \qquad (19)$$

where $\varphi = (\Delta\beta - \alpha_0)/(\alpha_0 - \Delta'\beta)$ with $\alpha_0^2 = k^2 - \omega^2/c^2$. α_0 is the decay coefficient in vacuum and must be positive. It needs to be emphasized that the existence of surface modes requires Re(ρ) >0 .Eqs. (18) and (19) will be applied to determine the bulk polariton bands and surface polaritons.

Numerical simulations based on FeF$_2$/TlBr will be performed with TMM. The reason is that their resonant frequencies lie in the far infrared range and are close to each other. The physical parameters here applied are $H_e = 533$kG , $4\pi M_0 = 7.04$kG , $H_a = 197$kG , $\varepsilon_a = 5.5$, $\eta = 5\times10^{-4}$, and $\omega_r = 498.8$kG ($\approx 52.45cm^{-1}$) for AF layers; $\varepsilon_l = 30.4$, $\varepsilon_h = 5.34$, $\omega_T = 48cm^{-1}$, $\xi = 8\times10^{-3}$ for IC layers. The external field $H_0 = 3.0T$ [10,33,34].

The MPP spectra are displayed in Fig.2, 3, and 5. In these spectra for dimensionless reduced f

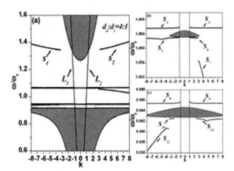

Figure 2. Bulk polariton bands (shaded areas) and surface polaritons (thick solid curves) for $d_a / d_i = 4:1$ via TMM: (a) a whole dispersion pattern; (b) and (c) are the partially enlarged figures. After Wang & Ta, 2010.

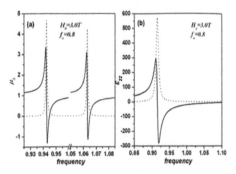

Figure 3. Effective magnetic permeability and dielectric function for $d_a / d_i = 4:1$. (a) The magnetic permeability and (b) the dielectric function. After Wang & Ta, 2010.

The MPP spectra are displayed in Fig.2, 3, and 5. In these spectra for dimensionless reduced frequency ω/ω_r versus k, the shaded regions stand for the bulk bands whose boundaries are determined by Eq. (18) or (19) with $Q = 0$ and $k = 0$, respectively. Meanwhile, the surface modes are obtained by Eqs. (18), (19) and (21). The thick curves with the serially numbered sign S denote the surface polaritons. The photonic lines $k^2 = (\omega / c)^2$ in vacuum are labeled as L₁ and L₂.Fig.2 shows the bulk bands and surface modes for the ratio $d_a / d_i = 4:1$ ($f_a = 0.8$), which are obtained by the TMM. Four bulk continua appear including two minibands (see Figs. 2(b) and 2(c)). The top and bottom bands (see Fig.2(a)) correspond to the positive real parts of the effective magnetic permeability and dielectric function, so the effective refraction index is positive in the two bands. However, the two mini bands (see Figs.2(b) and 2(c)) are related to the negative real parts of the permeability and dielectric function, which can be found from Fig.3. It means that the effective refraction index is negative in the mini bands. Ref.14 proved that the negative refraction and left-handedness exist in the mini bands, where the transmission and refraction of the same structure were examined in the absence of the external magnetic field.

Fig.4 displays the bulk bands and surface modes for ratio $d_a / d_i = 1:1$. The top bulk band is distinctly ascended to the high frequency direction, together with surface mode 1and 2. The bottom bulk band is widened conspicuously. Contrary to the previous situation, the two mini bands get significantly narrower. Compared with Fig.2, the slopes of surface modes 7 and 12 diminish appreciably, meaning their group velocity dwindles as the AF volume fraction (f_a) decreases.

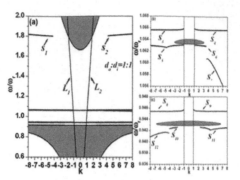

Figure 4. Bulk polariton bands and surface modes with the parameters and explanations as the same as those in Fig.2, except for $d_a / d_i = 1:1$. After Wang & Ta, 2010.

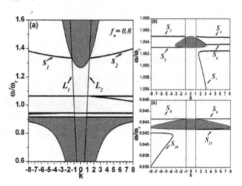

Figure 5. Bulk polariton bands and surface polaritons for $d_a / d_i = 4:1$. The interpretations of (a), (b) and (c) are the same as those of Fig.2. After Wang & Ta, 2010.

2.1.3. Limiting case of small period (EMM)

To examine the limiting case of small period or long wavelength is meaningful in physics. We let $d_a \to 0$ and $d_i \to 0$ in Eqs.(18),(19) and then find

$$Q^2 + (\mu_{xx}^e / \mu_{yy}^e)k^2 - \varepsilon_{zz}^e \mu_\lambda^e (\omega / c)^2 = 0, \qquad (20)$$

for the bulk modes with $\mu_\lambda^e = (\mu_{xx}^e \mu_{yy}^e - \mu_{xy}^{e2}) / \mu_{yy}^e$, and

$$\rho + (\mu_{xy}^e / \mu_{yy}^e)k + \alpha_0 \mu_\lambda^e = 0, \quad \rho^2 = (\mu_{xx}^e / \mu_{yy}^e)k^2 - \varepsilon_{zz}^e \mu_\lambda^e (\omega / c)^2 = 0, \qquad (21)$$

for the surface polaritons. If the external magnetic field implicitly included in Eqs.(20) and (21) is equal to zero, the dispersion relations can be reduced to those in our earlier paper [10].Hence equations (20) and (21) also can be considered as the results achieved by the EMM.

Fig.5 shows the bulk bands and surface modes for the ratio $d_a / d_i = 4:1$ ($f_a = 0.8$), which are obtained by the EMM. For the bulk bands, we see that the results obtained within the two methods almost are equal. However, the surface modes obtained by the EMM start from the photonic lines and are continuous, but those achieved by the TMM do not. It is because the interfacial effects are efficiently considered within the TMM, but the EMM neglects these. For the surface polaritons, their many main features attained by the two methods are still analogous. 12 surface mode branches are seen from Fig.2 within the TMM, but 11 surface modes from Fig.5 within the EMM. 10 surface mode branches arise in the common vicinities of two AF resonant frequencies and TO phonon frequency.Except branches 1 and 2, all surface modes are nonreciprocal and their non-reciprocity results from the magnetic contribution in AF layers. Surface polaritons 1 and 2 should be called the quasi-phonon polaritons since the contribution of the magnetic response to the polaritons is very weak in their frequency range. Another interesting feature is that many surface modes possess negative group velocities ($d\omega/dk < 0$), which is due to the combined contributions of magnetic damping and phonon damping.

2.2. MPPs in two-dimension AF/IC PCs

In this part, we consider such an AF/IC PCs constructed by periodically embedding cylinders of ionic crystal into an AF, as shown in Fig.6. We focus our attention on the situation where the external magnetic field and the AF anisotropy axis both are along the cylinder axis, or the z-axis. The surface of the MPC is parallel to the x-z plane. L and R indicate the lattice constant and cylindrical radius, respectively. We introduce the AF filling ratio, $f_a = 1 - \pi R^2 / L^2$, and the IC filling ratio, $f_i = \pi R^2 / L^2$. Our aim is to determine general characteristics of the surface and bulk polaritons with an effective-medium method under the condition of $\lambda \gg L$ (λ is the polariton wavelength).

2.2.1. EMM for the two-dimensional AF/IC PCs

When the AF/IC PCs cell size is much shorter than the wavelength of electromagnetic wave, an EMM can be established for one to obtain the effective permeability and permittivity of the AF/IC PCs. The principle of this method is in a cell, an electromagnetic-field component continuous at the interface is assumed to be equal in the two media and equal to the corresponding effective-field component in the MPC, but one component discontinuous at the interface is averaged in the two media into another corresponding effective-field component [30,33,40-41]. Because the interface between the two media is of cylinder-style, before establishing an EMM, a TMM should be introduced. This matrix is

$$T = \begin{pmatrix} \cos\theta & \sin\theta & 0 \\ -\sin\theta & \cos\theta & 0 \\ 0 & 0 & 1 \end{pmatrix}, \tag{22}$$

Thus, we find the expression of the permeability in the cylinder coordinate system

$$\bar{\mu}_a = T\bar{\mu}_a'T^{-1} = \begin{pmatrix} \mu_{rr} & \mu_{r\theta} & 0 \\ -\mu_{r\theta} & \mu_{\theta\theta} & 0 \\ 0 & 0 & 1 \end{pmatrix}. \tag{23}$$

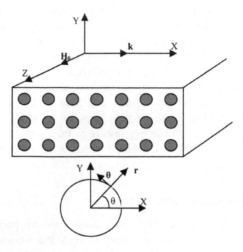

Figure 6. Geometry configuration and coordinate system.

with $\mu_{rr} = \mu_{\theta\theta} = \mu_{xx}$ and $\mu_{r\theta} = \mu_{xy}$. The theoretical processes of obtaining effective magnetic permeability, $\bar{\mu}^e$, and electric permittivity, $\bar{\varepsilon}^e$, are presented as follows. According to the principle, we can introduce the following equations:

$$H_z = h_{az} = h_{iz}, \; H_\theta = h_{a\theta} = h_{i\theta}, \tag{24}$$

$$B_r = b_{ar} = b_{ir}, \tag{25}$$

$$H_r = f_a h_{ar} + f_i h_{ir}, \tag{26}$$

$$B_\theta = f_a b_{a\theta} + f_i b_{i\theta}, \; B_z = f_a b_{az} + f_i b_{iz}, \tag{27}$$

where the field components on the left side of Eqs. (24)-(27) are defined as the effective components in the AF/IC PCs and those on the right side are the field components in the AF and IC media within the cell. In the AF, the relation between b and h is determined by (23) in the $r\theta z$ system, but in the IC, the relation is

$$\vec{b} = \vec{h}. \tag{28}$$

After defining the relation between the effective fields in the AF/IC PCs, $\vec{B} = \bar{\mu}^e \cdot \vec{H}$, the effective permeability resulting from (24)-(27) is

$$\bar{\mu}^e = \begin{pmatrix} \mu_{rr}^e & \mu_{r\theta}^e & 0 \\ -\mu_{r\theta}^e & \mu_{\theta\theta}^e & 0 \\ 0 & 0 & 1 \end{pmatrix}, \tag{29}$$

with

$$\mu_{rr}^e = \mu_{rr}/(f_a + \mu_{rr}f_i) \tag{30}$$

$$\mu_{\theta\theta}^e = f_a\mu_{rr} + f_i + f_a f_i \mu_{r\theta}^2/(f_a + \mu_{rr}f_i) \tag{31}$$

$$\mu_{r\theta}^e = \mu_{r\theta}f_a/(f_a + \mu_{rr}f_i) = -\mu_{\theta r}^e. \tag{32}$$

Formula (29) is the expression of the effective magnetic permeability in the $r\theta z$ system. When we discuss the surface polaritons, the AF/IC PCs will be considered as a semi-infinite system with a single plane surface, so the corresponding permeability in the rectangular coordinate system (orthe xyz system) will be used. Applying the transformation matrix (22) again, we find

$$\bar{\mu}_\perp^e = \begin{pmatrix} \mu_{rr}^e \cos^2\theta + \mu_{\theta\theta}^e \sin^2\theta & (\mu_{rr}^e - \mu_{\theta\theta}^e)\sin\theta\cos\theta + \mu_{r\theta}^e & 0 \\ (\mu_{rr}^e - \mu_{\theta\theta}^e)\sin\theta\cos\theta - \mu_{r\theta}^e & \mu_{rr}^e \sin^2\theta + \mu_{\theta\theta}^e \cos^2\theta & 0 \\ 0 & 0 & 1 \end{pmatrix}. \tag{33}$$

If one applies directly this form into the Maxwell equations, the resulting wave equation will be very difficult to solve. Thus, a further approximation is necessary. We think that if the wavelength of an electromagnetic wave is much longer than the cell size, then the wave will feel very slightly the structure information of the AF/IC PCs. Here, the averages of some physics quantities are important. Hence, $\bar{\mu}_\perp^e$ is averaged with respect to angle θ and one determines the averaged effective magnetic permeability,

$$\bar{\mu}_{\perp a}^e = \begin{pmatrix} (\mu_{rr}^e + \mu_{\theta\theta}^e)/2 & \mu_{r\theta}^e & 0 \\ -\mu_{r\theta}^e & (\mu_{rr}^e + \mu_{\theta\theta}^e)/2 & 0 \\ 0 & 0 & 1 \end{pmatrix}. \tag{34}$$

This means $\mu_{xx}^e = \mu_{yy}^e = \mu^e$. In physics, this AF/IC PCs should be gyromagnetic and be of C4-symmetry, as shown by Fig.1, which leads to the xx and yy elements of the final permeability should be equal and its xy element equal to -yx element.

By a similarprocedure, the effective dielectric permittivity can be easily found. According to the principle of EMM, we present the equations for the electric-field and electric-displacement components as follows,

$$E_z = \mathbf{e}_{az} = e_{iz}, \ E_\theta = \mathbf{e}_{a\theta} = e_{i\theta}, \ D_r = d_{ar} = d_{ir}, \tag{35}$$

$$D_Z = f_a d_{az} + f_i d_{iz}, \ D_\theta = f_a d_{a\theta} + f_i d_{i\theta}, \ E_r = f_a e_{ar} + f_i e_{ir}, \tag{36}$$

with $\vec{d}_{a(i)} = \varepsilon_{a(i)} \vec{e}$ in the AF or IC. After using the definition, $\vec{D} = \ddot{\varepsilon}^e \vec{E}$, the effective dielectric permittivity of the AF/IC PCs in the $r\theta z$ system is determined as

$$\ddot{\varepsilon}^e = \begin{pmatrix} \varepsilon_{rr}^e & 0 & 0 \\ 0 & \varepsilon_{\theta\theta}^e & 0 \\ 0 & 0 & \varepsilon_{zz}^e \end{pmatrix}, \tag{37}$$

With

$$\varepsilon_{rr}^e = \varepsilon_a \varepsilon_i / (f_a \varepsilon_i + f_i \varepsilon_a), \ \varepsilon_{\theta\theta}^e = f_a \varepsilon_a + f_i \varepsilon_i, \ \varepsilon_{zz}^e = f_a \varepsilon_a + f_i \varepsilon_i \tag{38}$$

Transforming (37) into the form for the xyz system, we see

$$\ddot{\varepsilon}_\perp^e = \begin{pmatrix} \varepsilon_{rr}^e \cos^2\theta + \varepsilon_{\theta\theta}^e \sin^2\theta & \varepsilon_{rr}^e \cos\theta\sin\theta - \varepsilon_{\theta\theta}^e \sin\theta\cos\theta & 0 \\ \varepsilon_{rr}^e \cos\theta\sin\theta - \varepsilon_{\theta\theta}^e \sin\theta\cos\theta & \varepsilon_{rr}^e \sin^2\theta + \varepsilon_{\theta\theta}^e \cos^2\theta & 0 \\ 0 & 0 & \varepsilon_{zz}^e \end{pmatrix}. \tag{39}$$

Then, its average value with respect to angle θ is

$$\ddot{\varepsilon}_{\perp a}^e = \begin{pmatrix} (\varepsilon_{rr}^e + \varepsilon_{\theta\theta}^e)/2 & 0 & 0 \\ 0 & (\varepsilon_{rr}^e + \varepsilon_{\theta\theta}^e)/2 & 0 \\ 0 & 0 & \varepsilon_{zz}^e \end{pmatrix}. \tag{40}$$

We see $\varepsilon_{xx}^e = \varepsilon_{yy}^e = \varepsilon^e$. Now, this AF/IC PCs can be considered as an effective medium with effective electric permittivity $\ddot{\varepsilon}_{\perp a}^e$ and magnetic permeability $\ddot{\mu}_{\perp a}^e$. If the AF medium is taken as FeF2 with its resonant frequency about $\omega_r / 2\pi c = 52.45 cm^{-1}$, proper wavelengths should be between 170 and 190μm. When the cell size is taken as μm order of magnitude, such as 5μm, the effective-medium theory is available and expressions (34) and (40) are reasonable.

2.2.2. Dispersion equations of surface and bulk MPP

The effective permittivity (40) and permeability (34) are applied to determine the dispersion equations of surface and bulk MPP in the AF/IC PCs. In the geometry of Fig.6, if the magnetic field of a plane electromagnetic wave is along the z-axis, the sublattice magnetizations in the AF do not couple with it, so the AF plays a role of an ordinary dielectric. Thus, we propose the electric fields of polariton waves in the AF/IC PCs are along the z-axis and the magnetic field is in the x-y plane. For a surface polariton, its electric field decaying with distance from the surface can be written as

$$\vec{E} = [E_0 \exp(-\alpha_0 y) \exp(ikx - i\omega t)]\vec{e}_z, (y > 0) \tag{41}$$

$$\vec{E} = [E_1 \exp(\alpha y) \exp(ikx - i\omega t)]\vec{e}_z, (y < 0) \tag{42}$$

in the AF/IC PCs, where k is the wave-vector component along the x-axis, but α and α_0 are the decay coefficients and are positive for the surface polariton. The corresponding magneticfield can be obtained with the relation $\nabla \times \vec{E} = i\omega \vec{B}$. From the Maxwell equations, we confirm the electric field obeys the wave equations

$$\nabla^2 \vec{E} + (\omega/c)^2 \vec{E} = 0, (y > 0) \tag{43}$$

$$\nabla^2 \vec{E} + (\omega/c)^2 \varepsilon^e_{zz} \mu^e_v \vec{E} = 0, (y < 0) \tag{44}$$

which lead to two relations

$$\alpha_0^2 = k^2 - (\omega/c)^2, \alpha^2 = k^2 - \varepsilon^e_{zz}\mu^e_v(\omega/c)^2 \tag{45}$$

with $\mu^e_v = [(\mu^e)^2 + (\mu^e_{r\theta})^2)]/\mu^e$. The dispersion relation can be found from the boundary conditions of the field components, E_z and H_x, continuous at the surface. Through a simple mathematical process, we obtain this relation

$$\alpha = -i\mu^e_{r\theta}k/\mu^e - \alpha_0 \mu^e_v, \tag{46}$$

where α and α_0 are determined by (45) with the conditions, $\alpha > 0$ and $\alpha_0 > 0$. Of course, it is very easy to find the dispersion relation of bulk polaritons. For the infinite AF/IC PCs, we find from the wave equation (44) that the dispersion relation in the x-y plane is

$$k^2 = \varepsilon^e_{zz}\mu^e_v(\omega/c)^2 \tag{47}$$

The bulk polariton bands are just such regions determined by (46). One can calculate directly the dispersion curves of the surface polariton from (45).

FeF2 and TlBr are utilized as constituent materials in the AF/IC PCs, which the parameters have been introduced in the last section. We place the AF/IC PCs into an external field of $H_0 = 5.0T$ along the z-axis and employ a dimensionless reduced frequency ω/ω_r in figures. Surface mode curves are plotted against wave vector k along the x-axis. Bulk modes form some continuous regions shown with shaded areas.

For comparison, we first present the polariton dispersion figures in the AF FeF2 and IC TlBr, as indicated in Fig. 7, respectively. For the AF, there exist three bulk bands and two surface modes. The surface modes appear in a nonreciprocal way and have a positive group velocity ($d\omega/dk > 0$). For the IC, its surface modes and bulk bands are depicted in Fig. 7(b). The surface modes are of reciprocity. Comparing Fig. 7(b) with the previous results without phonon damping [42], it is different that the two surface modes bear bended-back and end on the vacuum light line, due to the phonon damping.

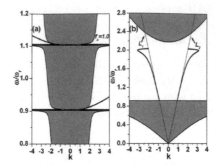

Figure 7. (a) Bulk polariton bands (shaded regions) and surface polaritons of FeF$_2$. (b) Bulk polariton bands and surface modes of TlBr, with two vacuum light lines (thin lines). After Wang & Ta, 2012.

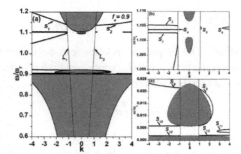

Figure 8. Bulk polariton bands (shaded regions) and surface polaritons of the MPC for $f_a = 0.9$: (a) a whole dispersion pattern;(b) and (c) are the partially amplified figures. After Wang & Ta, 2012.

For the AF/IC PCs with $f_a = 0.9$, Fig. 8 illustrates the dispersion features of magneto-phonon polaritons. Four bulk bands and 13 surface mode branches are found, where the surface modes are nonreciprocal (meaning the surface modes are changed when reversing their propagation directions). Two mini bulk bands and 11 surface mode branches exist in the vicinities of two AF resonant frequencies, where they are neither similar to those of the AF nor to those of the IC. Due to the combined contributions of the magnetic damping and phononic damping, the surface-mode group velocities become negative in some frequency ranges. For frequencies near the higher AF resonant frequency, the top bulk band bears a resemblance in nature to the top one of the AF, but the bottom band is analogous to the bottom one of the IC for frequencies close to the IC resonant frequency.

Figure 9 shows the bulk bands and surface modes for $f_a = 0.8$. The bulk bands in this figure are characteristically similar to those in Fig. 8, but the two mini bands are narrowed and the top one risesstrikingly. For the surface modes, mode 6 in Fig. 8 splits into two surface modes in Fig. 9. Modes 5 and 11 in Fig. 8 disappear from the field of view. The surface modes are still nonreciprocal.

The two mini bulk bands possess a special interest, corresponding to the negative effective magnetic permeability and negative effective dielectric permittivity of the AF/IC PCs. We present Fig. 10 for $f_a = 0.8$ to display the relevant effective permeability and permittivity. One can see that dielectric permittivity, ε_{zz}^e, is negative in a large frequency range. The two AF resonant frequencies lie in this range and the magnetic permeability, μ^e, is negative in the vicinities of the AF resonant frequencies. Thus, for electromagnetic waves traveling in the x-y plane, the refraction index is negative and the left-handedness can exist in the two mini bulk bands. When electromagnetic waves propagate along the z-axis, there is no coupling between AF magnetizations and electromagnetic fields, so the electromagnetic waves cannot enter this range where $\varepsilon^e < 0$ (see Eq. (1-30b)). The negative refraction and left-handedness were predicted in a one-dimension structure composed of identical materials [10].

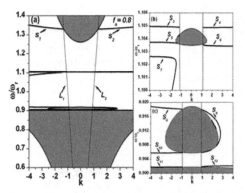

Figure 9. Bulk polariton bands and surface polaritons of the MPC for . The explanations of (a), (b) and (c) are identical to those of Fig.8. After Wang & Ta, 2012.

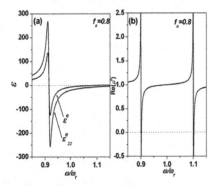

Figure 10. Effective permittivity and permeability in the MPC for : (a) the effective permittivity and (b) the effective permeability. After Wang & Ta, 2012.

3. Presence of left-handedness and negative refraction of AF/IC PCs

In the previous section, we have discussed MPPs in AF/IC PCs with the TMM and EMM for one- and two-dimension. Based on FeF2/TIBr, there are a number of surface and bulk polaritons in which the negative refraction and left-handedness can appear. In order to investigate the formation mechanism of LHM in AF/IC PCs, the external magnetic field and magnetic damping is set to be zero. In this case, according Eqs.(7) and (8), the effective permeability $\bar{\mu}^e$ and dielectric permittivity $\bar{\varepsilon}^e$ will be described as

$$\mu_{xx}^e = (f_a\mu + f_i), \ \mu_{yy}^e = \mu / (f_a + f_i\mu), \ \mu_{zz}^e = 1, \tag{48}$$

$$\varepsilon_{xx}^e = \varepsilon_{yy}^e = \varepsilon_a\varepsilon_i / (f_a\varepsilon_i + f_i\varepsilon_a), \ \varepsilon_{zz}^e = (f_a\varepsilon_a + f_i\varepsilon_i), \tag{49}$$

where $f_a = d_a / D$ is the AF filling ratio, and $f_i = d_i / D$ is IC filling ratio with $D = d_i + d_a$ as a bi-layer thickness.

Let us consider an incident plane electromagnetic wave propagating in the x-y plane as shown in Fig.11. Such a wave can be divided into two polarizations, a TE mode with its electric field parallel to axis z and a TM mode with its magnetic field parallel to axis z. According to Maxwell's equations, these wave vectors and frequencies of the two modes inside the film satisfy the following expressions

$$\frac{k_y^2}{\varepsilon_{zz}^e\mu_{xx}^e} + \frac{k_x^2}{\varepsilon_{zz}^e\mu_{yy}^e} = (\omega/c)^2 \ \ \left(\text{for TE mode}\right), \tag{50}$$

$$k_y^2 + k_x^2 = (\omega/c)^2 \varepsilon_{xx}^e\mu_{zz}^e \ \left(\text{for TE mode}\right), \tag{51}$$

Since $\mu_{zz}^e = 1$ is a positive real number, relation (51) corresponds to the case of an ordinary optical (when $\varepsilon_{xx}^e > 0$) or an opaque (when $\varepsilon_{xx}^e < 0$, contributing to an imaginary k_y) film, and so, we no longer consider the TM case, but deal with only the case of TE incident mode, and found the left-handed feature and negative refraction behavior. For the TE mode, the presence of left-handed feature (or negative refraction) needs the satisfaction of the prerequisite condition that $\varepsilon_{zz}^e < 0$ and μ_{xx}^e and μ_{yy}^e can not be simultaneously positive [i.e., at least one of μ_{xx}^e and μ_{yy}^e is negative, see (50)]. According to expressions (48) and (49),

$$\omega_T \le \omega \le \omega_T\sqrt{1 + f_i(\varepsilon_0 - \varepsilon_\infty)/(f_a\varepsilon_a + f_i\varepsilon_\infty)} \ (\text{for } \varepsilon_{zz}^e < 0), \tag{52}$$

$$\omega_r \le \omega \le \sqrt{\omega_r^2 + 2f_a\omega_a\omega_m} \ (\text{for } \mu_{xx}^e < 0), \tag{53}$$

$$\sqrt{\omega_r^2 + 2f_i\omega_a\omega_m} \le \omega \le \sqrt{\omega_r^2 + 2\omega_a\omega_m} \ (\text{for } \mu_{yy}^e < 0), \tag{54}$$

where $\omega_a = \gamma H_a$ and $\omega_m = 4\pi\gamma M_0$. Frequency region (52) is very large and covers regions (53) and (54) for the selected physical parameters. It is noted that both $\mu_{xx}^e < 0$ and $\mu_{yy}^e < 0$

can occur simultaneously only when AF layers are thicker than IC layers, which corresponds to spectral domain $(\omega_r^2 + 2f_i\omega_a\omega_m)^{1/2} \le \omega \le (\omega_r^2 + 2f_a\omega_a\omega_m)^{1/2}$. The wave electric field in the film can be written at

$$E_z = [A_0 \exp(ik_y y) + B_0 \exp(-ik_y y)]\exp(ik_x x - i\omega t), \tag{55}$$

and the corresponding magnetic field can be given by

$$\mathbf{H} = \frac{1}{i\omega}\left(\frac{\mathbf{e_x}}{\mu_{xx}^e}\frac{\partial E_z}{\partial y} - \frac{\mathbf{e_y}}{\mu_{yy}^e}\frac{\partial E_z}{\partial x}\right). \tag{56}$$

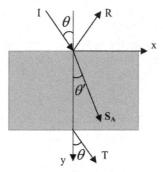

Figure 11. Draft for incidence, reflection, refraction and transmission rays. After Wang & Song, 2009.

The radiation in the film consists of two parts, one is the forward light (refraction light) related to amplitude A_0 and the other is the backward light (reflection light) related to amplitude B_0. Here, k_y is defined as a negative number, otherwise the refraction wave corresponds to amplitude B_0. These two situations are equivalent in essence. According to the definition of energy flow density of electromagnetic wave $\mathbf{S} = \mathrm{Re}(\mathbf{E} \times \mathbf{H}^*)/2$, the flow densities of the two lights can be given by

$$\mathbf{S_A} = \mathrm{Re}\left[\frac{|A_0|^2}{2\omega}(\frac{k_x}{\mu_{yy}^e}\mathbf{e_x} + \frac{k_y}{\mu_{xx}^e}\mathbf{e_y})\right], \mathbf{S_B} = \mathrm{Re}\left[\frac{|B_0|^2}{2\omega}(\frac{k_x}{\mu_{yy}^e}\mathbf{e_x} - \frac{k_y}{\mu_{xx}^e}\mathbf{e_y})\right]. \tag{57}$$

The inner product between a wave vector ($\mathbf{K_A} = [k_x, k_y, 0]$ or $\mathbf{K_B} = [k_x, -k_y, 0]$) and its corresponding energy flow is given by expression $I_A = \mathrm{Re}(\varepsilon_{zz}^e \omega |A_0|^2)/2$, or $I_B = \mathrm{Re}(\varepsilon_{zz}^e \omega |B_0|^2)/2$. Thus the angle between energy flow and wave vector can be expressed as

$$\alpha_j = \arccos[I_j/(|\mathbf{K_j}||\mathbf{S_j}|)], \tag{58}$$

where $j=A$ or B. It is obvious that α_A is equal to α_B and larger than $\pi/2$ in the range (52).

It can be seen from the expression (57) of S_A that its x component is negative and y component is positive when both $\mu^e_{xx} < 0$ and $\mu^e_{yy} < 0$, since k_x is positive and identical to the equivalent component of the incident wave vector. Thus an important condition is found for the existence of negative refraction, or AF layers must be thicker than IC layers. The refraction angle can be expressed as

$$\theta' = \arctan(k_x \mu^e_{xx} / k_y \mu^e_{yy}). \tag{59}$$

FeF$_2$ and TlBr are used as constituent materials where the AF resonant frequency ω_r is closer to the phonon resonant frequency ω_T and located in the far infrared regime. Fig.12 shows this angle as a function of frequency for $f_a = 0.8$ and 0.6. It can be seen from Fig.12 that for most of the frequency range occupied by the curves, angle α is at least bigger than 160° for various incident angles. So the wave vector, electric and magnetic fields form an approximate left-handed triplet. The operation frequency range becomes narrow as f_a decreases, as shown in Fig.12b.

As shown in Fig.13(a) for $f_a = 0.8$, the refraction angle is positive on the left side and negative on the right side of the intersection point of the curves. This corresponds to the following critical frequency obtained under the condition of $\mu^e_{xx} < 0$ and $\mu^e_{yy} < 0$:

$$\omega_c = \sqrt{\omega_r^2 + 2f_i \omega_m \omega_a}. \tag{60}$$

It can be seen from Fig.13(b) in comparing with Fig.13(a) that the frequency region of negative refraction is obviously narrower and the negative refraction angle becomes smaller. Numerical simulations also show both positive and negative refraction angles are in the spectral range of approximate left-handed feature shown in Fig.12.

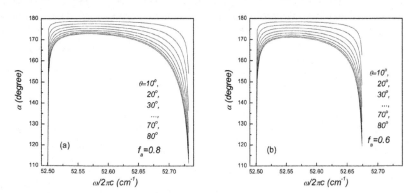

Figure 12. Angle α beteen refraction energy flow and corresponding wave vector versus frequency for different incident angles and for filling ratios (a) $f_a = 0.8$ and (b) $f_a = 0.6$. After Wang & Song, 2009.

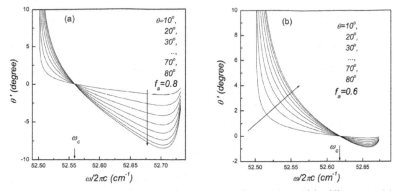

Figure 13. Refraction angle versus frequency for various incident angles, and for filling ratios (a) $f_a = 0.8$ and (b) $f_a = 0.6$. After Wang & Song, 2009.

4. Transmission, refraction and absorption properties of AF/IC PCs

In this section, we shall examine transmission, refraction and absorption of AF/IC PCs, where the condition of the period much smaller than the wavelength is not necessary. The transmission spectra based on FeF2/TIBr PCs reveal that there exist two intriguing guided modes in a wide stop band [11]. Additionally, FeF2/TIBr PCs possess either the negative refraction or the quasi left-handedness, or even simultaneously hold them at certain frequencies of two guided modes, which require both negative magnetic permeability of AF layers and negative permittivity of IC layers. The handedness and refraction properties of the system can be manipulated by modifying the external magnetic field which will determine the frequency regimes of the guided modes.

The geometry is shown in Fig. 1. We assume the electric field solutions in AF and IC layers as

$$\vec{E}_j = \vec{e}_z[A_j \exp(ik_j y) + B_j \exp(-ik_j y)]\exp(ik_x x - i\omega t), \qquad (61)$$

where $j = a, i$ signify AF or IC layers, respectively. The corresponding magnetic field solutions are also achieved via $\nabla \times \vec{E}_j = i\omega \vec{B}_j$. According to the boundary conditions of E_z and H_x continuous at interfaces, the relation between wave amplitudes in the two same layers of the two adjacent periods can be shown as a transfer matrix $\ddot{T}$ [43]. The matrix elements are expressed by the following equations:

$$T_{11} = \delta_a[(\Delta - \Delta')\cos(k_i d_i) + i(1 - \Delta\Delta')\sin(k_i d_i)] / (\Delta - \Delta'), \qquad (62)$$

$$T_{12} = i\delta_a^{-1}(1 - \Delta'^2)\sin(k_i d_i) / (\Delta - \Delta'), \qquad (63)$$

$$T_{21} = i\delta_a(\Delta^2 - 1)\sin(k_i d_i) / (\Delta - \Delta'), \qquad (64)$$

$$T_{22} = \delta_a^{-1}[(\Delta - \Delta')\cos(k_i d_i) + i(\Delta \Delta' - 1)\sin(k_i d_i)]/(\Delta - \Delta') \qquad (65)$$

with $\Delta = (ik_x \mu_\perp / \mu + k_a)/k_i \mu_v$ and $\Delta' = (ik_x \mu_\perp / \mu - k_a)/k_i \mu_v$. In AF layers, there are the relation $k_a^2 = \varepsilon_a \mu_v \omega^2/c^2 - k_x^2$ with $\mu_v = (\mu^2 - \mu_\perp^2)/\mu$, $k_x^2 = \omega^2/c^2 \sin^2\theta$ and ε_a the dielectric constant. The magnetic permeability tensor components of AF layers are represented by

$$\mu = 1 + \omega_m \omega_a \{1/[\omega_r^2 - (\omega_0 - \omega - i\sigma)^2] + 1/[\omega_r^2 - (\omega_0 + \omega + i\sigma)^2]\}, \qquad (66)$$

$$\mu_\perp = \omega_m \omega_a \{1/[\omega_r^2 - (\omega_0 - \omega - i\sigma)^2] - 1/[\omega_r^2 - (\omega_0 + \omega + i\sigma)^2]\}. \qquad (67)$$

with $\omega_m = 4\pi\gamma M_0$, $\omega_a = \gamma H_a$, $\omega_r = [\omega_a(2\omega_e + \omega_a)]^{1/2}$, $\omega_e = \gamma H_e$ and $\omega_0 = \gamma H_0$. M_0 represents the sublattice magnetization, and H_0, H_a and H_e indicate the external magnetic field, anisotropy and the exchange fields, respectively. ω_r denotes the zero-field resonant frequency. γ and σ are the gyromagnetic ratio and the magnetic damping. In IC layers, we have the relation $k_i^2 = \varepsilon_i \omega^2/c^2 - k_x^2$ with the dielectric function $\varepsilon_i = \varepsilon_h + \{(\varepsilon_l - \varepsilon_h)\omega_T^2\}/(\omega_T^2 - \omega^2 - i\xi\omega)$, where ε_h and ε_l are the high- and low-frequency dielectric constants, but ω_T indicates the frequency of the TO vibrating mode in the long-wavelength limition and ξ denotes the phonon damping. The magnetic permeability of IC layers is considered as $\mu_i = 1$. We assume here that the stacking number included in the magnetic superlattices is N. Then transmission and reflection coefficients of the AISL can be written as

$$\begin{pmatrix} 1 \\ r \end{pmatrix} = \frac{1}{2}\begin{pmatrix} 1+\Delta_1 & 1+\Delta_1' \\ 1-\Delta_1 & 1-\Delta_1' \end{pmatrix}\bar{T}^{N-1}\begin{pmatrix} k_i \delta_a \delta_i (1+\Delta)/(k_1 + k_i) & k_i \delta_a^{-1}\delta_i(1+\Delta')/(k_1 + k_i) \\ k_i \delta_a \delta_i^{-1}(1-\Delta)/(k_i - k_1) & k_i \delta_a^{-1}\delta_i^{-1}(1-\Delta')/(k_i - k_1) \end{pmatrix}\begin{pmatrix} t \\ t \end{pmatrix}. \qquad (68)$$

Note that incident wave amplitude is taken as 1. Therefore, the transmission and reflection coefficients can be determined with Eq. (68), and then the transmission ratio is $|t|^2$ and reflection ratio is $|r|^2$. Additionally, absorption ratio is represented with $A = 1 - |r|^2 - |t|^2$. Other quantities in Eq. (68) are $\Delta_1 = (ik_x \mu_\perp / \mu + k_a)/k_1 \mu_v$, $\Delta_1' = (ik_x \mu_\perp / \mu - k_a)/k_1 \mu_v$, $\delta_i = \exp(ik_i d_i)$, $k_1^2 = \omega^2/c^2 \cos^2\theta$.

As described in Ref. [8], magnetic superlattices possess two mini-bands with negative group velocity. When the incident wave is located in the frequency regions corresponding to the two mini-bands, what are the optical properties of the AF/IC PCs? In the preceding section, the expressions of transmission and absorption to be used have been derived. To grasp handedness and refraction properties of the AF/IC PCs, the refraction angle and propagation direction need to be determined. Therefore, subsequently we give the expression of the refraction angle. However, this structure possibly possesses a negative refraction, and generally the directions of the energy flow of electromagnetic wave and the wave vector misalign. We start with the definition of energy flow ($\vec{S} = \text{Re}[\vec{E} \times \vec{H}^*]/2$) with a view to achieving refraction properties. Based on wave solutions of the electric field and Maxwell equations, the magnetic field components of the forward-going wave in AF layers are shown as

$$H_{ax} = \frac{(k_a + ik_x\mu_\perp / \mu)}{\omega\mu_0\mu_v} A_a \exp(ik_a y + ik_x x - i\omega t). \tag{69}$$

$$H_{ay} = \frac{(ik_a\mu_\perp / \mu - k_x)}{\omega\mu_0\mu_v} A_a \exp(ik_a y + ik_x x - i\omega t). \tag{70}$$

The magnetic field components of the forward-going wave in the adjacent IC layers are

$$H_{ix} = \frac{k_i A_i}{\omega\mu_0} \exp(ik_i y + ik_x x - i\omega t), \; H_{iy} = -\frac{k_x A_i}{\omega\mu_0} \exp(ik_i y + ik_x x - i\omega t). \tag{71}$$

The amplitudes of two neighboring layers satisfy

$$\begin{pmatrix} A_i \\ B_i \end{pmatrix} = \begin{pmatrix} (1+\Delta_1)\delta_a & (1-\Delta_2)\delta_a^{-1} \\ (1-\Delta_1)\delta_a & (1+\Delta_2)\delta_a^{-1} \end{pmatrix} \begin{pmatrix} A_a \\ B_a \end{pmatrix} \tag{72}$$

According to boundary conditions, the electric and magnetic fields of every layer are acquired when the incident wave is known. Then the expressions of refraction energy flow in all layers are written as

$$\vec{S}_j = \text{Re}\left(\vec{E}_j \times \vec{H}_j^*\right)/2 - \text{Re}\left(-H_{jy}^* E_{jz}\vec{e}_x + E_{jz}H_{jx}^*\vec{e}_y\right)/2, \; \left(j=a,i\right). \tag{73}$$

What needs to be emphasized is that we here concentrate only on the refraction, so only the forward-going wave corresponding to the first term in Eq.(61) is considered and the backward-going wave is ignored. Owing to refraction angles being different in various layers, the refraction angle of the AF/IC PCs should be effective one. The angle between the energy flow and wave vector, and the refraction angle of the AF/IC PCs are defined as

$$\alpha = \text{arccos}[(\vec{K} \cdot \vec{S}) / (|\vec{K}||\vec{S}|)], \tag{74}$$

$$\theta' = \text{arctg}[(f_1 S_{ax} + f_2 S_{ix}) / (f_1 S_{ay} + f_2 S_{iy})], \tag{75}$$

with $\vec{K} = k_x\vec{e}_x + \left(f_1 k_a + f_2 k_i\right)\vec{e}_y$ and $\vec{S} = f_1\vec{S}_a + f_2\vec{S}_i$, where volume fractions are $f_1 = d_a/\left(d_a + d_i\right)$ and $f_2 = 1 - f_1$, respectively.

Numerical calculations based on FeF$_2$/TlBr PCs. We take the AF layer thickness $d_a = 4\mu m$ and the thickness of IC layers $d_i = 1\mu m$. The stacking number is $N = 9$. Figure 14 shows the transmission spectra with specific angles of the incidence in an external magnetic field $H_0 = 3T$. As illustrated in Fig. 14(a), the forbidden band ranges from $0.9\omega_r$ to $1.2\omega_r$

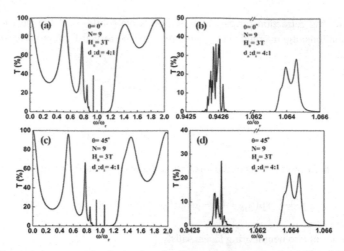

Figure 14. Transmission spectra for the fraction $f_1 = 0.8$, external magnetic field $H_0 = 3T$ and stacking number $N = 9$. (a) the incident angle $\theta = 0^0$; (b) a zoomed view of guided modes in (a); (c) the incident angle $\theta = 45^0$; (d) a zoomed view of guided modes in (c). After Wang & Ta, 2012.

corresponding to the band gap of magneto-phonon polariton in Ref. [8]. Here the most interesting may be that guided modes arise in the forbidden band. The two guided modes lie in the proximity of $\omega = 0.943\omega_r$ and $\omega = 1.064\omega_r$, corresponding to the mini-bands with negative group velocity in Ref. [8]. At the same time, the magnetic permeability of AF layers and the dielectric function for IC layers are both negative. To distinctly observe two guided modes, the partially enlarged Fig. 14(b) corresponding to Fig. 14(a) is exhibited. Seen from Fig. 14(b), the maximum transmission of the guided mode with lower-frequency is 40% and that of higher mode is 28.4%. As is well known, the optical thicknesses of films are determined by the frequency-dependent magnetic permeability and the dielectric function. Then the optical thicknesses of thin films are varied with the frequency of incident wave. The optical path of wave in media can also be altered by changing the incident angle. Fig. 14(c) shows the transmission spectrum with incident angle $45°$ and other parameters are the same as those in 14(a). The partially enlarged Fig. 14(d) corresponding to Fig. 14(c) is given. Compared with the normal incidence case, for $\theta \neq 0$ the forbidden band becomes wide and their maximum transmissions are reduced to 27.1% and 21.9%, but two guided modes keep their frequency positions unaltered.

As already noted, the damping is included and then the absorption appears. We are more interest in the two guided modes, so only the absorption corresponding to two guided modes will be considered in Fig. 15 (a) and (c) display the absorption spectra in the case of right incidence, but (b) and (d) illustrate the absorption spectra for incident angle $\theta = 45°$. We see that the absorption has a great influence on the transmission spectra. In the absorbing band at $\omega = 0.9426\omega_r$, relative tiny absorption corresponds exactly to the maximum transmission. In the absorbing band at $\omega = 1.024\omega_r$, the absorption is obviously strengthened with enhancing the incident angle.

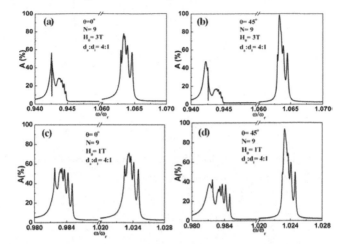

Figure 15. Partially enlarged absorption spectra with $f_1 = 0.8$, $N = 9$, $H_0 = 3T$ and $H_0 = 1T$. (a) corresponding to Fig. 14(b); (b) corresponding to Fig.14(d); (c) corresponding to Fig. 15(b); (d) corresponding to Fig. 15(d). After Wang & Ta, 2012.

To capture the handedness and refraction behaviors of the AF/IC PCs, the angle of refraction and the angle between the energy flow and wave vector are illustrated. Fig. 16 shows the angle α between the energy flow and wave vector of forward-going wave varies with frequency for $H_0 = 3T$ and $\theta = 45°$. As illustrated in Fig. 16, the angles in the vicinities of $\omega = 0.943\omega_r$ and $\omega = 1.064\omega_r$ are greater than $90°$, but less than $180°$. It indicates that the AF/IC PCs possesses a quasi left-handedness in these frequency regions.

Figure 17 shows the refraction angle θ' versus frequency under the same condition as Fig. 16. We find the refraction angles are negative in the neighborhood of $\omega = 0.943\omega_r$ and $\omega = 1.064\omega_r$. The two frequency ranges for negative angle do not completely coincide with those of the quasi-left-handedness in Fig. 17. Namely, the frequency regime of negative refraction near to $\omega = 0.943\omega_r$ is strikingly greater than that occupied by the quasi-left-handedness in Fig.17. However, the result is opposite in the vicinity of $\omega = 1.064\omega_r$. Therefore, we here reckon the left-handedness is not always accompanied by negative refraction. FeF$_2$/TlBrsuperlattices have the natures of either negative refraction or quasi left-handedness, or even simultaneously bear them at the certain frequencies of two guided modes.

To have a deeper understanding of the negative refraction and quasi- left-handedness of the AF/IC PCs, subsequently the expressions of the dielectric function

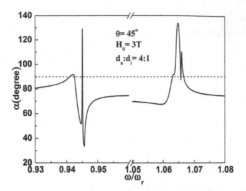

Figure 16. Angle α between energy flow and wave vector of down going wave versus the change of frequencies at $H_0 = 3T$, $\theta = 45^0$ and $f_1 = 0.8$. After Wang & Ta, 2012.

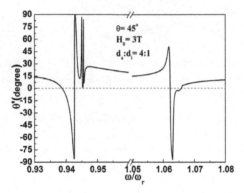

Figure 17. Refraction angle θ' versus the alteration of frequencies at $H_0 = 3T$, $\theta = 45^0$ and $f_1 = 0.8$. After Wang & Ta, 2012.

$\varepsilon_i = \varepsilon_h + \left\{(\varepsilon_l - \varepsilon_h)\omega_T^2\right\}/\left(\omega_T^2 - \omega^2 - i\xi\omega\right)$ of IC layers and the magnetic permeability of AF layers are analyzed. We find that when frequency lies in

$$\omega_T < \omega < \left(\varepsilon_l/\varepsilon_h\right)^{1/2}\omega_T, \tag{76}$$

the dielectric function ε_i is negative, namely the range $0.9152\omega_r < \omega < 2.1835\omega_r$. This completely covers the frequency range of AF resonance, so the dielectric function must be negative in the region of negative magnetic permeability μ_v. It is found from Fig.18 that the magnetic permeability μ_v is negative in certain regions, where the dielectric function ε_i is also negative. In our previous work [8], utilizing the effective medium theory, we verified the effective dielectric function and magnetic permeability are both negative in the long wavelength limit when μ_v and ε_i are negative. In other words, the AF/IC PCs is of negative refraction in the limit of long wavelength. Regarding the results arising from the two methods mentioned above, we conclude that the necessary condition of negative refraction or left-handedness is that μ_v and ε_i are both negative in this PCs.

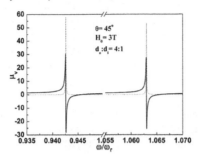

Figure 18. Voigt magnetic permeability μ_v versus the different frequencies at $H_0 = 3\,T$, $\theta = 45^0$ and $f_1 = 0.8$ (solid line denotes real parts and broken line indicates imaginary parts). After Wang & Ta, 2012.

5. Summary

This chapter aims to discover optical properties of AF/IC PCs in the presence of external static magnetic field. First, within the effective-medium theory, we investigated dispersion properties of MPPs in one- and two-dimension AF/IC PCs. The ATR (attenuated total reflection) technique should be powerful in probing these MPPs. Second, there is a frequency region where the negative refraction and the quasi left-handedness appear when the AF/IC PCs period is much shorter than the incident wavelength. Finally, an external magnetic field can be used to modulate the optical properties of the AF/IC PCs.

Author details

Shu-Fang Fu and Xuan-Zhang Wang
Key Laboratory for Photonic and Electronic Bandgap Materials, Ministry of Education, School of Physics and Electronic Engineering, Harbin Normal University, Harbin, China

Acknowledgement

This work was financially supported by the National Natural Science Foundation of China with Grant no.11084061, 11104050, and the Natural Science Foundation of Heilongjiang Province, with no. ZD200913.

6. References

[1] T. Goto, A .V. Baryshev, M. Inoue, A. V. Dorofeenko, A. M. Merzlikin, A. P. Vinogradov, A. A. Lisyansky, A. B. Granovsky, Tailoring surfaces of one-dimensional

magnetophotonic crystals: optical Tamm state and Faraday rotation, Phys. Rev. B. 79 (2009) 125103/1-5.

[2] R. H. Zhu, S. N. Fu, H. Y. Peng, Far infrared Faraday rotation effect in one-dimensional microcavity type magnetic photonic crystals, J. Magn. Magn. Mater. 323 (2011) 145-149.

[3] X. Z. Wang, The Faraday effect of an antiferromagnetic photonic crystal with a defect layer, J. Phys.:Condens. Matter. 17 (2005) 5447-5452.

[4] V. V. Pavlov, P. A. Usachev, R. V. Pisarev, D. A. Kurdyukov, S. F. Kaplan, A. V. Kimel, A. Kirilyuk, Th. Rasing, "Enhancement of optical and magneto-optical effects in three-dimensional opal/Fe3O4 magnetic photonic crystals," Appl. Phys. Lett. 93 (2008) 072502/1-3.

[5] V. V. Pavlov, P. A. Usachev, R. V. Pisarev, D. A. Kurdyukov, S. F. Kaplan, A. V. Kimel, A. Kirilyuk, Th. Rasing, "Optical study of three-dimensional magnetic photonic crystals opal/Fe3O4," J. Magn. Magn. Mater. 321 (2009) 840-842.

[6] R. Atkinson, "Limits of enhancement of the Faraday effect in ultra-thin magnetic layers by one-dimensional magnetophotonic crystals, "J. Phys. D: Appl. Phys. 39 (2006) 999-1005.

[7] M.M. Sigalas, C.M. Soukoulis, R. Biswas, K.M. Ho, "Effect of the magnetic permeability on photonic band gaps," Phys. Rev. B 56, (1997) 959-962.

[8] J. X. Ta, Y. L. Song, X. Z. Wang, "Magneto-phonon polaritons of antiferromagnetic/ion-crystal superlattices," J. Appl. Phys. 108 (2010) 013520/1-4.

[9] J. X. Ta, Y. L. Song, X. Z. Wang, "Magneto-phonon polaritons in two-dimension antiferromagnetic/ion-crystalic photonic crystals," Photonics and Nanostructures - Fundamentals and Applications 10(2012)/ 1-8.

[10] Y. L. Song, J. X. Ta, H. Li, X. Z. Wang, "Presence of left-handedness and negative refraction in antiferromagnetic/ionic-crystal multilayered films, " J. Appl. Phys. 106 (2009) 063119/1-4.

[11] J. X. Ta, Y. L. Song, X. Z. Wang, "Optical properties of antiferromagnetic/ion-crystal superlattices," J. Magn. Magn. Mater. 324 (2012) 72-74.

[12] J. B. Pendry, "Negative refraction makes a perfect lens," Phys. Rev. Lett. 85, (2000) 3966-3969.

[13] D. R. Smith and N. Kroll, "Negative refractive index in left-handed materials," Phys. Rev. Lett. 85, (2000) 2933-2936.

[14] D. Schurig and D. R. Smith, "Negative index lens aberrations," Phys. Rev. E 70, (2004) 065601-1-4.

[15] R. A. Shelby, D. R. Smith, and S. Schultz, "Experimental verification of a negative index of refraction," Science 292, (2001)77-79.

[16] D. R. Smith, W. J. Padilla, D. C. Vier, S. C. Nemat-Nasser, and S. Schultz, "Composite medium with simultaneously negative permeability and permittivity," Phys. Rev. Lett. 84, (2000) 4184-4187.

[17] W. T. Lu, J. B. Sokoloff, and S. Sridhar, "Refraction of electromagnetic energy for wave packets incident on a negative-index medium is always negative," Phys. Rev. E 69, (2004) 026604-1-5.

[18] C. G. Parazzoli, R. B. Greegor, K. Li, B. E. C. Koltenbah, and M. Tanielian, "Experimental verification and simulation of negative index of refraction using Snell's law," Phys. Rev. Lett. 90, (2003)107401-1-4.

[19] A. A. Houck, J. B. Brock, and I. L. Chuang, "Experimental observations of a left-handed material that obeys Snell's law," Phys. Rev. Lett. 90, (2003)137401-1-4.

[20] Zhaolin Lu, Janusz A. Murakowski, Christopher A. Schuetz, Shouyuan Shi, Garrett J. Schneider, and Dennis W. Prather, "Three-dimensional subwavelength imaging by a photonic-crystal flat lens using negative refraction at microwave frequencies," Phys. Rev. Lett. 95, (2005)153901-1-4.

[21] J. Lezec Henri, A. Dionne Jennifer, and A. Atwater Harry, "Negative refraction at visible frequencies," Science 316, (2007)430-432.

[22] A. Berrier, M. Mulot, M. Swillo, M. Qiu, L. Thylen, A. Talneau, and S. Anand, "Negative refraction at infrared wavelengths in a two-dimensional photonic crystal," Phys. Rev. Lett. 93, (2004) 073902-1-4.

[23] W. T. Lu and S. Sridhar, "Superlens imaging theory for anisotropic nanostructured metamaterials with broadband all-angle negative refraction," Phys. Rev. B 77, (2008) 233101-1-4.

[24] F. M. Wang, H. Liu, T. Li, Z. G. Dong, S. N. Zhu, and X. Zhang, "Metamaterial of rod pairs standing on gold plate and its negative refraction property in the far-infrared frequency regime," Phys. Rev. E 75, (2007) 016604-1-4.

[25] Andrea Alu and Nader Engheta, "Three-dimensional nanotransmission lines at optical frequencies: A recipe for broadband negative-refraction optical metamaterials," Phys. Rev. B 75, (2007) 024304-1-20.

[26] R.H. Tarkhanyana and D.G. Niarchos, "Wave refraction and backward magnon–Plasmon, polaritons in left-handed antiferromagnet/semiconductor superlattices," J. Magn. Magn. Materials, 312, (2007) 6-15.

[27] C. F. Klingshirn, Semiconductor Optics (Springer, Berlin 1997) Ch.13.

[28] A.D. Boardman, Electromagnetic Surface Modes, John Wiley, New York, 1982.

[29] R. T. Tagiyeva, M. Saglam,"Magnetic polaritons at the junctions of magnetic superlattice and magnetic material." Solid State Commun. 122 (2002) 413-417.

[30] N.Raj, D.R.Tilley, "Polariton and effective-medium theory of magnetic Superlattices ," Phys.Rev. B 36, (1987) 7003-07.

[31] Na Liu, Xiao-Dong Hou, Xiangdong Zhang, J. "Retarded modes in layered magnetic structures containing left-handed materials," Phys. Condens. Matter 20, (2008) 335210/1-5.

[32] C.A.A.Araújo, E.L.Albuquerque, D.H.A.L.Anselmo, M.S. Vasconcelos,"Magnetic polaritons in metamagnet layered structures: Spectra and localization properties," Phys. Lett. A 372, (2008) 1135-1140.

[33] Xuan-Zhang Wang, Tilley. D. R, "Retarded modes of a lateral antiferromagnetic /nonmagnetic superlattice," Phys. Rev. B 52, (1995) 13353-13357.

[34] M. C. Oliveros, N. S. Almeida, D. R. Tilley, J. Thomas, R. E. Camley, "Magnetostatic modes and polaritons in antiferromagnetic-non-magnetic superlattices," J. Phys. Condens. Matter 4, (1992) 8497-8510.

[35] I. Balin, N. Dahan, V. Kleiner, E. Hasman, "Slow surface phonon polaritons for sensing in the midinfrared spectrum," Appl. Phys. Lett. 94 , (2009) 111112/1-3.

[36] A. J. Huber, B. Deutsch, L. Novotny, R. Hillenbrand, "Focusing of surface phonon polaritons," Appl. Phys. Lett. 92 , (2008) 203104/1-3.

[37] K. Abraha, D.E. Brown, T. Dumelow, T.J. Parker, D.R. Tilley, "Obliqu-incidence far-infrared reflectivity study of the uniaxial antiferromagnet FeF2," Phys. Rev. B 50 ,(1994) 6808-6816.

[38] E.D. Palik, Handbook of Optical Constants, Vol.3, Academic Press, San Diego, 1998, pp.926; R.C. Fang, Solid State Spectroscopy, Press., Univ. of Sci. and Tech. of China, Hefei, 2001, pp. 234 (in Chinese).

[39] Xuan-Zhang Wang and Hua Li, Nonlinear propagation of electromagnetic waves in antiferromagnets, in Propagation of Electromagnetic Waves in Complex Matter, edited by Ahmed Kishk, (Intech-Open Access Publisher, 2011)pp.55-96, Ch.3

[40] X.Z. Wang, D.R. Tilley, "Magnetostatic surface and guided modes of lateral-magnetic-superlattice films," Phys. Rev. B 50, (1994) 13472-13479.

[41] X.Z. Wang, X.R. Xu, "Nonlinear Magnetostatic surface waves of magnetic multilayers: Effective-medium theory," Phys. Rev. B 63, (2001) 054415/1-11.

[42] D.N. Mirlin, in: V.M. Agranovich, D.L. Mills (Eds.), Surface phonon polaritons in dielectrics and semiconductors, in Surface Polaritons, North-Holland Press, Amsterdam, 1982, pp. 5-67.

[43] X. F. Zhou, J. J. Wang, X .Z. Wang, D.R. Tilley, "Reflection and transmission by magnetic multilayers," J. Magn. Magn. Mater. 212, (2000) 82-90.

Permissions

The contributors of this book come from diverse backgrounds, making this book a truly international effort. This book will bring forth new frontiers with its revolutionizing research information and detailed analysis of the nascent developments around the world.

We would like to thank Dr. Orhan Yalçin, for lending his expertise to make the book truly unique. He has played a crucial role in the development of this book. Without his invaluable contribution this book wouldn't have been possible. He has made vital efforts to compile up to date information on the varied aspects of this subject to make this book a valuable addition to the collection of many professionals and students.

This book was conceptualized with the vision of imparting up-to-date information and advanced data in this field. To ensure the same, a matchless editorial board was set up. Every individual on the board went through rigorous rounds of assessment to prove their worth. After which they invested a large part of their time researching and compiling the most relevant data for our readers. Conferences and sessions were held from time to time between the editorial board and the contributing authors to present the data in the most comprehensible form. The editorial team has worked tirelessly to provide valuable and valid information to help people across the globe.

Every chapter published in this book has been scrutinized by our experts. Their significance has been extensively debated. The topics covered herein carry significant findings which will fuel the growth of the discipline. They may even be implemented as practical applications or may be referred to as a beginning point for another development. Chapters in this book were first published by InTech; hereby published with permission under the Creative Commons Attribution License or equivalent.

The editorial board has been involved in producing this book since its inception. They have spent rigorous hours researching and exploring the diverse topics which have resulted in the successful publishing of this book. They have passed on their knowledge of decades through this book. To expedite this challenging task, the publisher supported the team at every step. A small team of assistant editors was also appointed to further simplify the editing procedure and attain best results for the readers.

Our editorial team has been hand-picked from every corner of the world. Their multi-ethnicity adds dynamic inputs to the discussions which result in innovative outcomes. These outcomes are then further discussed with the researchers and contributors who give their valuable feedback and opinion regarding the same. The feedback is then collaborated with the researches and they are edited in a comprehensive manner to aid the understanding of the subject.

Apart from the editorial board, the designing team has also invested a significant amount of their time in understanding the subject and creating the most relevant covers. They scrutinized every image to scout for the most suitable representation of the subject and create an appropriate cover for the book.

The publishing team has been involved in this book since its early stages. They were actively engaged in every process, be it collecting the data, connecting with the contributors or procuring relevant information. The team has been an ardent support to the editorial, designing and production team. Their endless efforts to recruit the best for this project, has resulted in the accomplishment of this book. They are a veteran in the field of academics and their pool of knowledge is as vast as their experience in printing. Their expertise and guidance has proved useful at every step. Their uncompromising quality standards have made this book an exceptional effort. Their encouragement from time to time has been an inspiration for everyone.

The publisher and the editorial board hope that this book will prove to be a valuable piece of knowledge for researchers, students, practitioners and scholars across the globe.

List of Contributors

Orhan Yalçın
Niğde University, Niğde, Turkey

Manish Sharma, Sachin Pathak and Monika Sharma
Indian Institute of Technology Delhi, India

R.Singh and S.Saipriya
School of Physics, University of Hyderabad, Central University P.O., Hyderabad, India

Chi-Kuen Lo
Department of Physics, National Taiwan Normal University, Taipei, Taiwan

H. Montiel
Centro de Ciencias Aplicadas y Desarrollo Tecnológico, Universidad Nacional Autónoma de México, Del. Coyoacán, México DF 04510, México

G. Alvarez
Escuela Superior de Física y Matemáticas del IPN, U.P.A.L.M, Edificio 9, San Pedro Zacatenco, México DF 07738, México

Gabriela Vázquez-Victorio, Ulises Acevedo-Salas and Raúl Valenzuela
Institute for Materials Research, National Autonomous University of México, México

Stefania Widuch
Center for Magnetism and Magnetic Nanostructures, University of Colorado, Colorado Springs, Colorado, USA
School of Physics (M013), University of Western Australia (UWA), Crawley, WA, Australia
Institute of Physics, University of Silesia, Katowice, Poland

Robert L. Stamps
School of Physics and Astronomy, Kelvin Building, University of Glasgow, Scotland, UK

Danuta Skrzypek
Institute of Physics, University of Silesia, Katowice, Poland

Zbigniew Celinski
Center for Magnetism and Magnetic Nanostructures, University of Colorado, Colorado Springs, Colorado, USA

Marie Yoshikiyo, Asuka Namai and Shin-ichi Ohkoshi
Department of Chemistry, School of Science, The University of Tokyo, Tokyo, Japan

Shin-ichi Ohkoshi
CREST, JST, K's Gobancho, 7 Gobancho, Chiyoda-ku, Tokyo, Japan

Shu-Fang Fu and Xuan-Zhang Wang
Key Laboratory for Photonic and Electronic Bandgap Materials, Ministry of Education, School of Physics and Electronic Engineering, Harbin Normal University, Harbin, China

CPSIA information can be obtained
at www.ICGtesting.com
Printed in the USA
BVOW07*0332230118
505966BV00002B/10/P